To my wife
for putting up with it

James Edward Gordon was born in 1913. He
took a degree in naval architecture at Glasgow
University and worked in wood and steel
shipyards, intending to design sailing ships. On the
outbreak of the Second World War he moved to
the Royal Aircraft Establishment at Farnborough,
where he worked on wooden aircraft, plastics and
unorthodox materials of all kinds. He designed the
sailing rescue dinghies carried at one time by most
bomber aircraft. He later became head of the
plastics structures section at Farnborough and
developed a method of construction in
asbestos-reinforced plastics which is now used for
a number of purposes in aircraft and rockets. For
several frustrating years he worked in industry on
the strength of glass and the growth of strong
'whisker' crystals. In 1962 he returned to
government service as superintendent of
an experimental branch at Waltham Abbey
concerned with research and development of
entirely new structural materials – most of which
are based on 'whiskers'. He was Industrial Fellow
Commoner at Churchill College, Cambridge and
is now Professor of Materials Technology at the
University of Reading. He has been awarded the
British Silver Medal of the Royal Aeronautical
Society for work on aircraft plastics and also the
Griffith Medal of the Materials Science Club for
contributions to material science. His book,
Structures or Why Things Don't Fall Down, has also
been published in Penguins.

Professor J. E. Gordon is married and has two
grown-up sons. His spare-time interests are sailing,
Greek, photography and skiing.

J. E. Gordon

The New Science of Strong Materials

or Why you don't fall through the floor

SECOND EDITION

Penguin Books

Penguin Books Ltd, Harmondsworth,
Middlesex, England
Penguin Books, 625 Madison Avenue,
New York, New York 10022, U.S.A.
Penguin Books Australia Ltd, Ringwood,
Victoria, Australia
Penguin Books Canada Ltd, 2801 John Street,
Markham, Ontario, Canada L3R 1B4
Penguin Books (N.Z.) Ltd, 182–190 Wairau Road,
Auckland 10, New Zealand

First published 1968
Reprinted with minor revisions 1971, 1973, 1974
Reprinted 1975
Second edition 1976
Reprinted 1977, 1978, 1979
First published in the United States by
Walker & Co., New York, 1968

Made and printed in Great Britain by
Hazell Watson & Viney Ltd, Aylesbury, Bucks
Set in Monotype Times Roman

'*The habit of apprehending a technology in its completeness: this is the essence of technological humanism, and this is what we should expect education in higher technology to achieve. I believe it could be achieved by making specialist studies the core around which are grouped liberal studies which are relevant to these specialist studies. But they must be relevant; the path to culture should be through a man's specialism, not by-passing it . . .*

'*A student who can weave his technology into the fabric of society can claim to have a liberal education; a student who cannot weave his technology into the fabric of society cannot claim even to be a good technologist.*'

Lord Ashby, *Technology and the Academics*

Contents

List of Plates 11
Acknowledgements 12
Foreword to First Edition 13
Foreword to Second Edition 14

1 The new science of strong materials – *or how to ask awkward questions* 15

Part One Elasticity and the theory of strength 25

2 Stresses and strains – *or why you don't fall through the floor* 27

3 Cohesion – *or how strong ought materials to be?* 63

4 Cracks and dislocations – *or why things are weak* 77

Part Two The non-metallic tradition 99

5 Crack-stopping – *or how to be tough* 101

6 Timber and cellulose – *or Wooden ships and Iron men* 129

7 Glue and plywood – *or mice in the gliders* 154

8 Composite materials – *or how to make bricks with straw* 173

Part Three The metallic tradition 207

9 Ductility in metals – *or the intimate life of the dislocation* 209

10 Iron and steel – *Hephaistos among the Satanic Mills* 231

11 The materials of the future – *or how to have second thoughts* 254

Appendix 1 On the various kinds of solids – *and what about treacle?* 269

Note on Conversion of Units 275

Appendix 2 Simple beam formulae – *or do your own stressing* 276

Suggestions for further study 280

Index 283

List of Plates

1 Wells Cathedral (By courtesy of A. F. Kersting)

2 Cantilever beam loaded at one end

3 S. S. Schenectady

4 Silica glass rod bent elastically to a strain of
$7\frac{1}{2}$ per cent (By courtesy of Dr J. W. Morley
and Messrs Rolls Royce Ltd)

5 Cracks caused by slight accidental contact on
the surface of Pyrex glass (*Proceedings of the
Royal Society*, A.249)

6 Deliberate scrape on a microscope cover-glass
(*Proceedings of the Royal Society*, A.249)

7 Devitrification in silica glass (*Proceedings of
the Royal Society*, A.249)

8 The cellular structure of wood greatly enlarged
(Dr John Dinwoodie, Crown Copyright)

9 Tin whiskers growing spontaneously on a
tin-plated surface (By courtesy of C. C. Evans)

10 Growth steps on large whisker crystal
(By courtesy of C. C. Evans)

11 Effect of weak interfaces in stopping or hindering
cracks (By courtesy of G. Cooper)

12 Whiskers or needle crystals of hydroquinone
growing from solution in water

13 Fracture in thin polythene sheet (By courtesy of
Professor Sir Charles Frank, F.R.S.)

14 The first direct photograph of an edge
dislocation (By courtesy of Sir James Menter,
F.R.S.)

15 Trevithick's 1805 high-pressure locomotive
(Science Museum, London)

16 Hedley's eight-wheeled locomotive (Science
Museum, London)

17 Blenkinsop's rack locomotive (Science
    Museum, London)

Acknowledgements

We are grateful to the Royal Society for permission
to use the following figures: Chapter 2, Figure 5;
Chapter 3, Figure 1; Chapter 4, Figure 2; Chapter 5,
Figures 3, 4 and 5; Chapter 9, Figure 3. Figure 2,
Chapter 2, is reproduced by courtesy of J. T. Norton
and B. M. Loring (*Welding Journal, Research
Supplement, June* 1941).

Foreword to First Edition

Anyone venturing to write a book covering nearly the whole field of strong materials must give many hostages to his colleagues for his mistakes, his oversimplifications, his omissions and for his pure ignorance. Such a book is necessarily highly selective and, in practice, the choice must be personal – I hope that I may be forgiven for writing about what has interested me. There are people much better fitted than I to write, for instance, about alloy steels or titanium.

Although the book is published by permission of the Ministry of Technology, the views expressed, like the errors, are entirely my own.

Materials science and elasticity are generally pursued as rather mathematical subjects. However, I have cut out the whole of the mathematics except for a very little genuinely childish elementary algebra which can be followed by anybody with a negligible effort. In most cases I have substituted Roman titles for the Greek ones which are commonly used by the professionals (i.e. s for σ (stress), e for ε (strain) and so on). Simple transliteration returns these formulae to their common text-book form.

I should like to thank Dr W. D. Biggs of Christ's College, Cambridge, for reading the book in proof and for making useful suggestions and Mr Gerald Leach for encouragement and patient literary criticism.

Foreword to Second Edition

I owe a considerable debt of gratitude to many of my colleagues at Reading University for their help in revising this book and bringing it up to date; especially to Professor W. D. Biggs, Dr Richard Chaplin and Dr Giorgio Jeronimidis. I have also taken into account comments and criticisms from correspondents in various parts of the world who have been kind enough to write to me.

With regard to units, it is impossible to please everybody. I have tried to achieve some kind of compromise or symbiosis between S.I. and traditional Anglo-Saxon which I hope that readers will not find too emotionally disturbing.

Chapter 1   The new science of strong materials

or how to ask awkward questions

> '*Now, how curiously our ideas expand by watching
> these conditions of the attraction of cohesion! – how
> many new phenomena it gives us beyond those of the
> attraction of gravitation! See how it gives us great
> strength. The things we deal with in building up the
> structures on the earth are of strength (we use iron,
> stone and other things of great strength); and only
> think that all those structures you have about you –
> think of the "Great Eastern", if you please, which is
> of such size and power as to be almost more than man
> can manage – are the result of cohesion and attrac-
> tion.*'
>
> Michael Faraday (1791–1867), *On the various
> forces of Nature.*

Why do things break? Why do materials have any strength at
all? Why are some solids stronger than others? Why is steel
tough and why is glass brittle? Why does wood split? What do
we really *mean* by 'strength' and 'toughness' and 'brittleness'?
Are materials as strong as we ought to expect them to be? How
far can we improve existing types of materials? Could we make
altogether different kinds of materials which would be much
stronger? If so, how, and what would they be like? If we really
could make better materials then how and where should we make
use of them?

Towards the end of his life Faraday was beginning to ask some
of these questions but he could make no progress in answering
them and, indeed, it is only quite recently that we have been able
to do so. Yet in merely asking questions about cohesion Faraday
was very much ahead of his time and for many years afterwards
strength and cohesion remained very unfashionable subjects with
the scientific establishment.

This book is about how we have come to understand the
strength of materials, about how the strengths of metals and
wood and ceramics and glass and bone are interrelated and about

how and why these materials behave in various kinds of structures such as machinery and tools and ships and aeroplanes and buildings and bridges.

Because what we can achieve technically has always been limited by the weaknesses of the materials of construction this new science is important. Instead of accepting our materials as something provided, arbitrarily, by Providence – as people used to until very recently – we can understand why they behave as they do and moreover, we can see much more clearly how they might be modified and improved. As a consequence, we are beginning to see our way to making radically better materials, unlike anything which has existed before, and these may open up quite new possibilities to the engineers.

Metals and non-metals

'Gold is for the mistress – silver for the maid –
Copper for the craftsman cunning at his trade.'
'Good!' said the Baron, sitting in his hall,
'But Iron – Cold Iron – is master of them all.'

Rudyard Kipling, 'Cold Iron'.

'I'll make ships out of wood; and this time I'll enjoy life and be happy.'

Weston Martyr, *The Southseaman.*

The great division in technology has always been that between metals and non-metals, and, although the masters like Brunel were well skilled in using both, the majority of engineers are committed to one tradition or the other. The division arises because the properties of metals and non-metals are clearly so very different – we shall presently see why – and thus the ways of using them have to be quite different. However, I am inclined to think that the mental processes involved may also be in some way opposed; metallurgists and metal engineers are apt to be practical down-to-earth people who stand no nonsense, but the non-metallurgists are probably more lyrical and imaginative.

Although perhaps we must not make too much of it, some such division has run right through the history of technology; in this book we shall review the progress of these two traditional branches of engineering in the light of modern materials science and try to see what the real problems were and why changes came when they did.

The cheapening and improvement of iron and steel during the eighteenth and nineteenth centuries* was the most important event of its kind in history – or perhaps just the most important event in history. In any case, iron and steel were particularly suited to Victorian artifacts and thus our contemporary technology is largely metallic. Metals, however, do not have a monopoly of strength. Some of the best combinations of lightness and strength are afforded by non-metals and the strongest substances in existence are the recently discovered 'whisker' crystals of carbon and of ceramics.

As the subject is developing, it now seems very possible that the coming new engineering materials will resemble much improved versions of wood and bone more closely than they will the metals with which most contemporary engineers are familiar. In the long run it is quite possible that this may affect the whole scale and character of our industries and, although it is unlikely that we shall go back to a William Morris world of woodcarvers and village carpenters or that metals will be superseded in the foreseeable future, it does make a study of the history of the use of strong materials as a whole – both metals and non-metals – relevant. Although the new techniques will be very sophisticated in many ways perhaps we may be able to get back to the patient humility of the craftsman in the face of his material which has got lost somewhere in our arrogant factories. This might result in more satisfying employment and perhaps in less industrial ugliness. If this is so, then the gain in human happiness will be very great.

We shall therefore use the modern scientific views on the strength of materials to illuminate the nature, the history and some of the applications of those materials of construction which seem to be the most important socially and technically. The selection of subjects is admittedly arbitrary and I have left out important materials, such as aluminium, where they do not seem

---

*The price of steel was reduced more than tenfold during the reign of Queen Victoria.

to illustrate any very interesting principle – *l'art d'ennuyer consiste à tout dire*.

The nature of materials science

It is clear that the strength of even the largest engineering structure depends in part upon chemical and physical events happening upon a molecular scale and so we shall not only have to let our ideas range freely up and down the scale of physical dimensions from the very big to the very small, but we shall also have to jump backwards and forwards from the ideas of chemistry to those of engineering. In the current phrase materials science is 'interdisciplinary'.

As soon as we start thinking about the mechanical properties of solids it becomes clear that, while we have some idea of 'how' materials behave, we have really very little idea of 'why'; naturally it is the 'why' questions which are generally the more sophisticated and the harder to answer. However, before we can tackle the reason for the way things behave we must be able to describe that behaviour accurately and objectively; this is the business of engineers. The man in the street may have rather vague views about how the solids around him deflect and break but engineers have to be precise about it and they have spent a good many generations in refining their descriptions and making them more objective. It is quite true that engineers had no idea at all why a piece of steel behaved in the way it did while a piece of concrete behaved quite differently but, at any rate, they described and measured these behaviours and wrote it all down in unreadable books. Armed with a knowledge of 'the properties' of their materials they were usually able to predict the behaviour of complicated structures – though, of course, they were sometimes wrong, in which case bridges fell down, ships sank or aeroplanes crashed.

This descriptive wisdom is embodied in the science of elasticity which defines the conditions under which structural materials receive, transmit and resist their loads; it will be necessary to have some understanding of it in order to see what the problems of strength are about. Without all the mathematics, the important principles of elasticity are really very simple but curiously difficult to understand. I think this is because most of us have grown to man's estate making use of some kind of instinctive knowledge about the strength of solids – if we hadn't we should have broken things and hurt ourselves even oftener than we did – and we think that we understand the subject intuitively and do not need telling. In fact, the real difficulty lies, not in learning about elementary elasticity, but in first getting rid of our preconceptions.

Anyone troubled with doubts on such matters is recommended to try to describe, objectively, the mechanical differences between chalk and cheese. On the whole, engineers can do this and, further, if we should wish to build a structure from either of these materials, engineers are able to predict the manner of its collapse. For the *reasons* for the differences between chalk and cheese, however, we must call on some of the other traditional divisions of science.

Solids are held together by the chemical and physical bonds between their atoms and molecules and any solid can be destroyed in several different ways, for instance, by mechanical fracture, by melting or by chemical attack. Since similar bonds have to be loosened in every case one might suppose that all these forms of dissolution were interrelated in some fairly simple way and that, now that chemists and physicists know so much about the nature of the bonds between atoms, they would have no special difficulty in explaining strength and other mechanical properties; in fact that fracture would have become practically a branch of chemistry.

As we shall see, strength is related, as of course it must be, to chemical bonding but the connexion is a roundabout one and cannot be deduced simply and entirely from classical chemistry and physics. It has turned out that not only do we need to interpret classical chemistry and physics by means of classical elasticity, but we also need to make use of rather new but important concepts such as dislocations and stress concentrations. In their day, these concepts were resisted by many of the orthodox.

It is undeniable that, until lately, the strength of materials as a science has lagged behind apparently more difficult, but perhaps more glamorous subjects and for a long time far more was known about things like wireless or the internal constitution of the stars than about what went on inside a piece of steel. In my opinion this is not so much because of the absolute difficulty of the subject but rather because of the difficulty of getting enough people in different disciplines to communicate and to take the subject seriously.

Chemists rather like to explain all the properties of matter in chemical terms and, when they had unscrambled the difficulties caused by the fact that chemists and engineers use different units to measure such things as energy, they found that their strength predictions were not only frequently a thousandfold in error but bore no consistent relationship with experiment at all. After this they were inclined to give the whole thing up and to claim that the subject was of no interest or importance anyway. Physicists did not take quite this attitude but, for a long time, most of them had other fish to fry – such as what goes on inside an atom.

Nowadays, of course, by an alliance between the physicists and the metallurgists, what occurs within a piece of metal is revealed in almost embarrassing detail, but for a long time classical metallurgy remained a descriptive science. Metallurgists knew that if one added this or that element to an alloy its properties would be affected. In the same way they knew that heating or cooling or hammering metals changed their mechanical behaviour. They could cut the metal open and observe differences in the gross crystal structure under the optical microscope but, although these differences were correlated with the behaviour of the metal, they could not be said in any way to 'explain' its behaviour at that level of explanation which we have become accustomed to demand.

Superstition and craftsmanship

Since the subject has proved so troublesome to scientists it was not to be expected that our ancestors would approach it in a very logical way and, in fact, no technical subject has been so deeply infested with superstition. A long and mostly gruesome book could, and perhaps should, be written about the superstitions associated with the making and fabrication of materials. In ancient Babylon the making of glass required the use of human embryos; Japanese swords were said to have been quenched by plunging them, red-hot, into the bodies of living prisoners. Cases of burying victims in the foundations of buildings and bridges were common – in Roman times a doll was substituted. All this was more or less in line with a good deal of primitive anthropology and seems to centre on the idea that the new structure should have a life of its own.

Latterly we have become less cruel but perhaps not much less superstitious. At any rate *some* element of irrationality about materials lingers in us all. For instance, the questions of old versus new, natural versus synthetic materials are ones which many people approach with an emotional fervour which is seldom based on real knowledge or experimental evidence. These prejudices are strongest in the non-structural applications where there is 'nothing like wool' or 'nothing like leather' but they also spread over into the structural field.

All these attitudes really amount to the idea of a kind of vitalism in materials, a 'vis-viva' on which the reliability of the substance depends; a workman will tell you that such and such has broken 'because the nature has gone out of it'. During the last war I was concerned with the supply of bamboos for making kites for anti-aircraft barrages. An importer of bamboos told me that he found it difficult to stock the lengths which we needed because they took up so much space since they had to be stored horizontally. I asked him why he did not store them vertically. 'If I did that,' he said, 'the nature would run out of the ends.'

In the past, of course, instinct and experience were the only guides to the choice of materials and to the design of structures and devices. The best traditional craftsmen were sometimes fairly good but it is a mistake to exaggerate the virtues of traditional design; the workmanship may have been excellent but the engineering design is often mediocre and sometimes shockingly bad. The wheels really did keep coming off coaches because coach-builders were not clever enough to attach them properly. In the same way wooden ships have always leaked, quite unnecessarily, in a sea-way because shipwrights didn't understand the nature of a shearing stress and I am afraid that many of them still don't.

This excursion into the pre-scientific side of the subject might seem out of place in a book devoted to the modern science of materials, but the science of materials, like the science of medicine, has had to make its way in the teeth of a great many traditional practices and old wives' tales. Not to take account of the pit of anti-knowledge from which materials science has had to extricate itself would be unrealistic.

Atoms, chemistry and units of measurement

Even though the connexions between the strength of materials and classical physics and chemistry are not always simple or direct the subject does, of course, ultimately rest firmly on foundations of basic chemistry and physics and, for those who may have forgotten some of their 'O' level science, there is an appendix at the end of the book which tries to recap very briefly the basic minimum of physics and chemistry upon which the arguments are based. However, in the understanding of materials science it may be that an apprehension of the dimensions and scale of the various phenomena is as important as knowing the rules of chemistry and physics. In other words the 'laws' of science provide the rules of the game but the dimensions of the chess-boards – the scales upon which the games of nature and technology are played out – vary almost unimaginably. It is therefore worth spending a few moments over questions of scale and units of measurement.

Lord Kelvin used to say that one could not be said to know anything about a phenomenon until one could measure something about it and to do so naturally requires units of measurement. Latterly, S.I. units have been introduced in England, in the schools and elsewhere, but to talk entirely in terms of Newtons and metres is probably to bring in an extra difficulty or stumbling-block for most ordinary people in English-speaking countries who generally continue to think in pounds and tons and feet and inches, and so, for the larger measurements, we shall use both English and S.I. units side by side. When we come to very small measurements, however, we can all become metric and perhaps more rational. Since materials science is dealing to a considerable extent with the very small, these very small units, which are not in everyday household use, are important.

I MICRON (1μm) is $\frac{1}{10,000}$ of a centimetre, that is $\frac{1}{1,000}$ of a millimetre. The smallest thing that one can see in the ordinary way with the naked eye is roughly $\frac{1}{10}$ of a millimetre, that is about 100 microns across. The smallest thing one can see with an ordinary optical microscope is usually a little less than half a micron across. Actually the smallest thing one can see is a good deal affected by the lighting conditions. In a beam of strong light in a dark room one can see dust particles with the naked eye about 10 microns across or even less. Because one micron is around the limit of resolution in optical microscopes it is a favourite unit with biologists and other users of the optical microscope.

I ÅNGSTRÖM UNIT (1Å) is $\frac{1}{10,000}$ of a micron, that is $\frac{1}{100,000,000}$ of a centimetre. (10Å = 1 nanometre (nm) = 10^{-9} metre)

These are favourite units with electron microscopists and they are the units used for measuring atoms and molecules. The newer electron microscopes can see – as rather woolly blobs – particles about five Ångströms across, that is about a thousand times smaller than the best optical microscopes can achieve.* Here again the resolution is a good deal governed by the viewing conditions.

Atoms are what all matter is made of. Atoms themselves consist of a very small and heavy nucleus surrounded by a large or small cloud of planetary electrons which are waves, particles or negative charges of electricity and are very small indeed. The whole affair varies a good deal in weight and size according to the kind of atom but may be thought of as a hard but fuzzy ball very roughly two Ångströms in diameter. This is inconceivably small by ordinary standards and it is quite impossible that we should ever see individual atoms by ordinary visible light – though obviously we see them in the mass when we look at any solid.

It may be worth emphasizing that the smallest particle one can see with the naked eye is about 500,000 atoms across and the

---

* But then electron microscopes cost nearly a hundred times as much as optical microscopes.

smallest particle one can see with the optical microscope is about 2,000 atoms across. With the electron microscope one can see arrays of atoms in crystals, like soldiers on parade, quite easily and with a device called the field emission microscope one can see individual atoms – at least one can see that there is something which looks like a sheep in a fog on a dark evening. However if the microscope resolution were much better, as it may perhaps become, this merely raises the rather metaphysical question of what one would expect to 'see' anyway. Nothing very concrete surely?

Note There is a conversion table between English and S.I. units on page 275.

Part One

Elasticity and the theory of strength

Chapter 2   Stresses and strains

or why you don't fall through the floor

> *'It had been his custom to engage Wan in philosophical discussion at the close of each day and on this occasion he was contrasting the system of Ka-ping, who maintained that the world was suspended from a powerful fibrous rope, with that of Tai-u who contended that it was supported upon a substantial bamboo pole. With the clear insight of an original and discerning mind Ah-shoo had already detected the fundamental weakness of both theories.'*

Ernest Bramah, *Kai Lung unrolls his mat.*

We are so used to not falling through the floor that we never stop to think why we don't. However the problem of how any inanimate solid is able to resist a load at all worried both Galileo (1564–1642) and Hooke (1635–1702). The understanding of simple structures and how they resist loads is a good example of a problem which, except in its molecular aspects, requires no sophisticated apparatus and could, in theory, be solved almost entirely by pure reason. This is not to say that the subject is easy; it is intellectually very difficult. The genius of Galileo and Hooke lay as much in recognizing that an important problem existed as in their contributions towards solving it, significant as these were.

As a matter of fact the general problem was probably beyond the scientific potential of the seventeenth century and it was not until well into the nineteenth that any reasonably complete idea of what was happening in a structure existed; even then this knowledge was confined to a few rather despised theoreticians. For a long time 'practical' engineers went on as they always had done, by rule of thumb. It took a long history of controversy and a series of disasters like the Tay bridge to convince these people of the usefulness of proper strength calculations.* Also it was found that reliable calculations enabled structures to be made

---

* It is said that at one time railway bridges were collapsing in the United States at the rate of twenty-five a year.

more cheaply because one could more safely economize in material. Nowadays the main difference between the qualified professional engineer on the one hand and the bench mechanic and the do-it-yourself amateur on the other lies not so much in mechanical ingenuity and skill as in an understanding of the problems of strength and energy.

Let us begin at the beginning with Newton (1642–1727) who said that action and reaction are equal and opposite. This means that every push must be matched and balanced by an equal and opposite push. It does not matter how the push arises. It may be a 'dead' load for instance: that is to say a stationary weight of some kind. If I weigh 200 pounds and stand on the floor, then the soles of my feet push downwards on the floor with a push or thrust of 200 pounds (or 900 Newtons, if you must); that is the business of feet. At the same time the floor must push upwards on my feet with a thrust of 200 pounds (or 900 Newtons); that is the business of floors. If the floor is rotten and cannot furnish a thrust of 200 pounds then I shall fall through the floor. If, however, by some miracle, the floor produced a larger thrust than my feet have called upon it to produce, say 201 pounds, then the result would be still more surprising because, of course, I should become airborne. Similarly, if a chair weighs 50 pounds, then the floor obliges by producing an upward force of exactly the 50 pounds which are needed to support the chair in its accustomed station in life. On the other hand, the force need not be a stationary weight. If I drive my car into a wall, the wall will respond by producing exactly enough force to stop the car at whatever speed it may be going, even if it kills me. Again, the wind, blowing where it listeth, pushes on my chimney pots but the chimney pots, bless them, push back at the wind just as hard, and that is why they don't fall off.

All this is merely a restatement of Newton's third law of motion which says, roughly speaking, that if the *status quo* is to be maintained then all the forces on an object must cancel each other out. This law does not say anything about how these various forces are generated. As far as the applied loads are concerned, the manner of their generation is usually straightforward: the weight of a 'dead' load arises from the action of the earth's gravitation upon the mass of the load and in the case of stopping a moving load (whether a solid, a liquid or a gas) the forces generated are those needed to decelerate the moving mass (Newton's second law of motion). The business of all structures is the conservative one of maintaining the *status quo* and in order to do this they must somehow generate adequate forces to oppose the loads which they have to carry. We can see how a weight presses down on the floor but how does the floor press up on the weight?

The answer to this question is far from obvious and the problem was the more difficult for Galileo and Hooke, in the early days of scientific thought, because the biological analogy is confusing and the tendency is, or was, to begin thinking about a problem in an anthropomorphic way. An animal has really two mechanisms for resisting loads. Its inert parts – bones, teeth and hair – resist by just the same means as any other inert solid but the living animal as a whole behaves in a quite different manner. People and other animals resist mechanical forces by pushing back in an active way: they tense their muscles and push or pull as the situation may require. If I stretch out my hand and you put a weight on it such as a pint of beer, then I have to increase the tensions in certain muscles so as to sustain the load. I am enabled to do this because the tensions in our muscles can be continually adjusted by an elaborate biological mechanism. However, the maintenance of biological tensions requires the continual expenditure of actual work (like driving a car fitted with a fluid flywheel while it is hard up against a wall – the engine is working away and using petrol and the car is pushing against the wall but neither the car nor the wall are moving). For this reason my arm muscles will sooner or later get tired and so I shall have to drink the beer to relieve them. One remains standing, not like a tripod standing inertly on the ground, but by a series of deliberate, though perhaps unconscious, adjustments of the body muscles. One gets tired standing up, and, if the muscular processes are interrupted by fainting or death, there is a dramatic collapse.

In an inanimate solid these living processes are not available. Structural materials are passive and cannot push back deliberately, so that they do not, in the ordinary sense, get tired.

They can only resist outside forces *when they are deflected*; that is, they must give way to the load to a greater or less extent in order to generate any resistance at all. By 'deflection', in this context, we do not mean that the solid moves bodily, as a whole and without changing its shape, but rather that the geometrical form of the solid is to some extent distorted so that some parts of it at least become shorter or longer by stretching or contracting within themselves. There is, and there can be, no such thing as a truly rigid material. Everything 'gives' to some extent and, as we have said, the realization that this is what structural engineering is about is what divides the professional from the amateur engineer.

When I climb a tree the deflections of the boughs under my weight will probably be very large, perhaps a matter of several inches, and are easily seen. However, when I walk across a bridge the deflections may be imperceptibly small. These are only questions of degree: there is always some deflection. Unless the deflections under loads are excessively large for the purpose of the structure they are not a fault but an inborn and unavoidable characteristic of structures with which it is the business of this chapter to come to terms. Most of us have sat in an aeroplane and watched the wing-tips going up and down. This is quite all right; the designer meant them to be like that.

It is probably obvious by this time that these deflections, be they large or small, generate the forces of resistance which make a solid hard and stiff and resistant to external loads. In other words, a solid deflects exactly far enough to build up forces which just counter the external load applied to it. This is the automatic process at the basis of all structures.

How are these forces generated? The atoms in a solid are held together by chemical forces or bonds (see Appendix 1) which may perhaps be thought of as electrical springs since there is nothing 'solid' in any crude sense to make any other kind of spring. It is these forces which bind solids together and also make the rules of chemistry. There is no distinction between the chemical bonds between atoms whose fracture yields the energy of gunpowder or petrol, and the chemical bonds which make steel and rubber strong and elastic.

When a solid is altogether free from mechanical loads (which, strictly speaking, is very seldom) these chemical bonds or springs are in their neutral or relaxed position (Figure 1). Any attempt

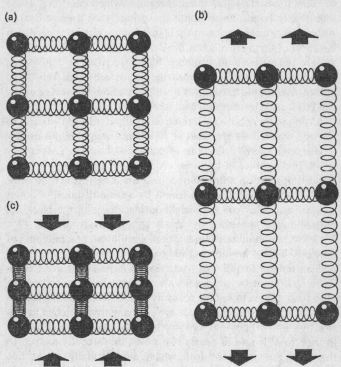

Figure 1. Simplified model of distortion of interatomic bonds under mechanical strain.

    (a) Neutral, relaxed or strain-free position.

    (b) Material strained in tension, atoms further apart, material gets longer.

    (c) Material strained in compression, atoms closer together, material gets shorter.

to push them closer together (which we call compression) or to stretch them further apart (which we call tension) involves shortening or lengthening the interatomic springs, by however little, throughout the material. This is because the middle part of the atoms may be regarded as rigid and, furthermore, in a solid the atoms do not generally exchange places – at least at moderate or 'safe' loads. The only thing, therefore, which can 'give' is the interatomic bond. These bonds or springs vary a good deal in stiffness or springiness (or, as the layman might put it, in 'strength') but most of them are much stiffer than the metal springs to which we are accustomed in ordinary life. From this it follows, of course, that the forces between the atoms are often very large indeed. We should expect this if we think about the forces which can be released by chemical explosives and fuels.

Although there is no such thing as a truly rigid solid – that is to say one which does not yield at all when a weight is put on it – in everyday life the deflections of common objects are often very small. For instance, if I take an ordinary builder's ceramic brick, stand it upright on a firm surface and tread on it, then the brick will be compressed along its length by a total distance of about $\frac{1}{50,000}$ of an inch. Any two neighbouring atoms in the brick are pushed nearer together by about $\frac{1}{500,000}$ Ångström unit (2×10^{-14}cm., or about one hundredth of a millionth of a millionth of an inch). This is an inconceivably small distance but a perfectly real movement for all that. Actually, in large structures the deflections are not by any means always tiny. In order to support their load, that is to say the roadway and the cars, the suspension cables of the Forth road bridge are permanently stretched in tension by about 0·1 per cent, or something like ten feet (three metres) in their total length of nearly two miles, or three kilometres. In this case the atoms of iron which are normally about two Ångström units apart when at rest and unloaded are kept about $\frac{2}{1,000}$ Ångström units further apart than they would be in the unstressed state.

That atoms really do move further apart when a material is stretched has been checked experimentally many times and by different methods. The most obvious way is by X-ray diffraction of stretched and unstretched specimens. The standard way of measuring the distance between the atoms in a crystal is to study the way in which an X-ray beam is deflected when it passes through the crystal. This method has been used now for sixty years or more and it is nowadays capable of considerable accuracy. It is found that the atoms in a metal, for instance, move apart or together exactly in proportion to the amount by which the metal as a whole is stretched or compressed. Changes in inter-atomic spacing up to about 1·0 per cent have been observed. Some actual measurements up to about 0·3 per cent are shown in Figure 2.

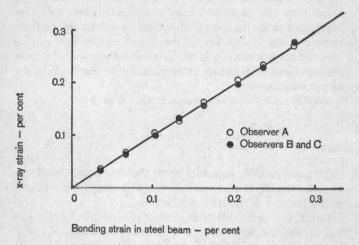

Figure 2. Comparison of strains determined by X-rays (two-exposure method) and strains computed from the curvature of a bent beam. Annealed mild steel.

What are stresses and strains, and why bother?

All this brings us to the question of stresses and strains, words which the layman is apt to regard as alarming, distressing and confusing. This is perhaps partly because the words may conjure up the idea of a wilderness of mathematics but probably more because the words have been borrowed or stolen by non-scien-tists to describe the mental condition of human beings. In this connotation the words have no very precise meaning and com-monly stress and strain are used interchangeably as if they meant the same thing. All this is a pity because in science the two words have quite simple, clear and distinct meanings.

So far we have thought, as much as we have thought about it at all, of the force acting on a material as being the total load upon it. This might be any weight, and we have thought of the deflection under that load as being the total deflection, whatever the dimensions of the object, large or small. This is all very well but it gives us no proper standard of comparison between a big object under a big load and a small object under a little load. One might want to use the same kind of steel for a tiny part in a typewriter and also for the keel of an aircraft carrier: how can we compare its performance in the two jobs? Until we have some proper objective standards of comparison we cannot take the subject much further.

Stress is simply load per unit area. That is to say:

$$s = \frac{P}{A} \text{ (where } s = \text{stress}, P = \text{load}, A = \text{area.)}$$

This may possibly look frightening, but it is exactly analogous to such everyday remarks as 'the cost of butter is 60p a pound' or 'my car does thirty miles to the gallon'.

Hence, to revert to the brick, if its cross section is 3 inches by 4 inches then its end has an area of 12 square inches and, if I tread on it with a weight of 200 pounds, the compressive stress which I cause in the brick is clearly:

$$s = \frac{P}{A} = \frac{200}{12} = 16\tfrac{2}{3} \text{ pounds on each square inch, or pounds per square inch, or lb./in.}^2, \text{ or p.s.i.}$$

Similarly if the brickwork pier of a bridge has a cross-section measuring 20 feet by 5 feet and it is crossed by a railway engine weighing 100 tons (224,000 pounds) then the compressive stress in the brickwork will be roughly 16 p.s.i. We can say with confidence therefore that in both cases the stress in the bricks is similar and, if one structure is safe, so most probably will be the other. As far as the bricks are concerned, their molecules are being pushed together with an identical force although the engine is ponderous and I am relatively small. This is obviously the sort of thing that engineers want to know.

In English-speaking countries stresses are traditionally expressed in pounds per square inch or tons per square inch. Continental engineers generally use kilogrammes per square centimetre. With S.I. the use of Newtons per square metre (N/m²) usually produces embarrassingly large numbers and so we generally use Meganewtons per square metre (MN/m²); (1 Meganewton = one million Newtons). In this book we shall use p.s.i. and MN/m² side by side.* Nevertheless we must be clear that we are applying the concept to the conditions at any cross-section or at a point and not especially to a square inch or to a square metre. Because the price of butter is 60p a pound this price does not apply especially to one pound; it is just as applicable to larger or smaller quantities.

Strain is just as simple: *Strain is the amount of stretch under load per unit length.*

Obviously, different lengths of material stretch different distances under the same load. So:

$$e = \frac{l}{L}$$

where e = strain, l = total amount of stretch, L = original total length.

So, if a rod 100 inches long stretches one inch under load, then it is subject to a strain of 1/100 or 0·01 or 1·0 per cent. So also is a rod 50 inches long which stretches ½ inch, and so on. It does not matter how fat or thin the rod is or what is causing the extension. We are only concerned with how much the component atoms and molecules are stretched apart, and so strain is again, like stress, independent of the size of the specimen. Strain is a *fraction* of the original length and so it remains just a fraction or a ratio (in other words a number) and has no units, British, S.I. or anything else.

* For conversion:
  1 MN/m² = 10·2 Kg/cm² = 146 p.s.i.
  1 p.s.i. = 0·00685 MN/m² = 0·07 Kg/cm²
  1 kg/cm² = 0·098 MN/m² = 14·2 p.s.i.

Hooke's law

The first man to grasp what was happening when an inert solid was loaded was Robert Hooke* who, besides being a physicist, was a notable architect and engineer and used to discuss the behaviour of springs and pendulums with the great clockmaker Thomas Tompion† (1639–1713). Hooke, or course, knew nothing about the chemical and electrical forces between atoms, but he realized that a 'spring' as a clockmaker might think of it is only a special case of the behaviour of any elastic solid and that there is no such thing as a truly rigid material, springiness being a property of every structure and of every solid.

Hooke, like Horace, did not suffer unduly from modesty and he staked his claim to priority in a number of fields by publishing in 1676 *A decimate of the centesme of the inventions I intend to publish* among which was 'The true theory of elasticity or springiness'. This heading was followed simply by the anagram 'ceiiinosssttuu'. The scientific public were left to make what they could of this until, in 1678, Hooke published *De potentia restitutiva, or of a spring* where the anagram was revealed as 'Ut tensio‡ sic uis' – 'As the extension, so the force'.

In other words, stress is proportional to strain and vice versa. So, if an elastic body such as a wire is stretched one inch under a load of 100 pounds it will stretch two inches under 200 pounds and so on, *pro rata*. This is known as Hooke's law and is regarded as one of the pillars of engineering.

As a matter of fact Hooke's law is really an approximation which arises from the character of the forces between atoms. There are several kinds of chemical bonds between atoms (see Appendix 1) but they all result in interatomic force curves which are similar in general shape (Figure 3). At very large strains – 5 to 10 per cent or so – stress is anything but proportionate to strain.

* For Hooke in general, see *Robert Hooke* by Margaret 'Espinasse.
† Tompion is the joy of modern sale-rooms where his lovely clocks fetch enormous prices.
‡ Tensio means, generally, not tension but extension in Latin. The truth seems to be that the Romans muddled up the two ideas. Literary writers probably never thought about the matter at all.

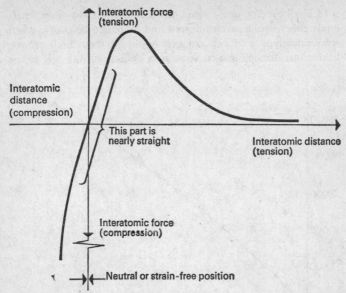

Figure 3. Relationship between the distance between two atoms and force between them.

However, in practical engineering materials, strains nearly always lie in the range ± 1·0 per cent either side of the neutral or strain-free position and for this range the relation between stress and strain is pretty well a straight line. Furthermore for small strains the whole process of extension and recovery is reversible and can usually be repeated many thousands or millions of times with identical results; the hair-spring of a watch which is coiled and uncoiled 18,000 times each hour is a familiar example. This type of behaviour by solids under loading is called 'elastic' and is widespread. Elastic behaviour, which is shown by the majority of engineering materials, contrasts with 'plastic' behaviour, shown to the extreme by putty and Plasticine, where the material does not obey Hooke's law in the initial loading and does not recover properly when the load is removed. The word 'elastic' is not especially dedicated to indiarubber and sock-suspenders and the science of elasticity is the study of stresses and strains in solids.

In Hooke's day, and indeed down to the last few years, materials either broke or else flowed and ceased to be elastic when strains much over 1·0 per cent were applied to them. So the shape of the interatomic force curve at large deflections was only of the

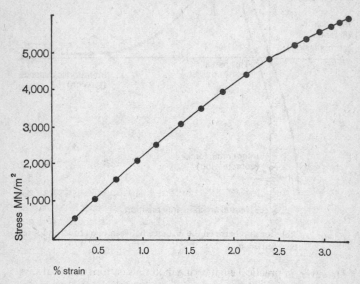

Figure 4. Stress-strain relationship for a very strong silicon whisker. This whisker or needle-like crystal was strained to 3·6 per cent in a testing machine and although the behaviour is 'elastic' it does not obey Hooke's law at the higher strains, the top of the graph being distinctly curved. This is because the interatomic force relationship is also curved at the higher strains. Other strong filaments, such as iron whiskers, have similar non-linear stress-strain curves at high stresses.

most academic interest because such stresses were never reached. Fairly recently, in the writer's laboratory and elsewhere, it has been possible to take very strong 'whisker' crystals up to strains between 3 and 6 per cent and the measurements confirm that Hooke's law is not literally true. The stress-strain curve bends over to follow the interatomic force curve which is derived from considerations, not of engineering, but of theoretical physics.

Figure 4 shows such a curve for a silicon whisker strained to over 3 per cent.

Young's modulus

Hooke stated that the deflections of springs and other elastic bodies were proportionate to the load which is applied to them but, of course, with different structures, the actual deflection under any given load will depend both upon the geometrical size and shape of the structure and also upon the material from which it is made. It is not clear how far Hooke distinguished elasticity as a property of a *material* from elasticity as a function of the *shape and dimensions* of the structure. We can get similar load-extension curves from a straight piece of rubber and from a helical piece of steel which we call a spring – this has always been a fruitful source of confusion. Certainly for something like a century after Hooke's time a state of intellectual muddle seems to have invested the few people who thought about elasticity and no clear distinction seems to have been made between these ideas.

Around 1800 Thomas Young (1773–1829) realized that, if we consider the stresses and strains in the material rather than the gross deflections of the structure, then Hooke's law can be written:

$$\frac{\text{stress}}{\text{strain}} = \frac{s}{e} = \text{constant}$$

Furthermore, Young realized that there was here a constant peculiarly characteristic of each chemical substance which, as he might have said, represents its 'springiness'. We call this constant 'Young's modulus' or E. There is no mystery about the word 'modulus', it just means a figure which describes a property of a material. Thus:

$$E = \frac{s}{e} = \frac{\text{stress}}{\text{strain}}$$

E therefore describes the elastic flexibility of a material as such; the flexibility of any given object will thus depend both upon the Young's modulus of the material from which it is made and also upon its geometrical shape.

It is said of Young that he was 'a man of great learning but unfortunately he never even began to realize the limitations of
comprehension of ordinary minds'.* Young published the idea of
his modulus in a rather incomprehensible paper in 1807 after he
had been dismissed from his lectureship at the Royal Institution
for not being sufficiently practical. Thus perhaps the most famous
and the most useful of all concepts in engineering, which defines
the stiffness or floppiness of a material, was not generally understood or absorbed into engineering practice until after Young's
death. Young's modulus is often called 'stiffness' in casual engineering conversation and will sometimes be called stiffness and
sometimes E in this book.

E is enormously important in engineering for two reasons.
First, we need to know with accuracy the deflections in a structure, as a whole and in its various parts, when it is loaded. A
moment's thought about bridges or aeroplanes or crankshafts
will show that this is so (Figure 5). Things must still fit together,

Figure 5.   Aircraft with strain of 1·6 per cent in wing spar booms.
$$\left(\text{Bend radius of beam} = \frac{\text{thickness}}{2 \times \text{strain}}\right)$$

or have the proper clearances, when the load is on.† A knowledge
of the E of the material being used is the first thing we need to
know in making these calculations. Secondly, although the lay-

* S. B. Hamilton, *History of Technology*, vol. 4, chapter 15.
† I once did a design study of a plastic railway carriage for British
Rail. One of the troubles was that, if the doors fitted properly when the
carriage was empty, they would neither open nor close when the carriage was
full of passengers in the rush hour.

man might suppose, as the early engineers seem to have done, that the stiffness of all common structural materials were very similar ('Well, it's stiff, isn't it, you can't *see* any deflections'), this is in fact very far from being the case and we not only need to know the *E*s of various materials such as wood and steel in order to calculate their deflections, but we must also arrange that the deflections of differing materials in a structure are compatible and that they share the load in the way we want them to.

Since, if we divide stress by a ratio – that is by a number without dimensions* – we must still have a stress, Young's modulus is therefore a stress in pounds per square inch, or what you will. It is that stress which would in theory double the length of a specimen, if it did not break first. One can also regard it as the stress to produce 100 per cent strain. As it will easily be imagined, the actual figure is likely to be a high one, usually at least a hundred times larger than the breaking stress of the material, because, as we have said, materials are apt to fracture in the ordinary way at 1 per cent elastic strain or less. The Young's modulus of steel, for example, is about 30,000,000 pounds per square inch. As we have also said, *E* varies very much according to the kind of chemical substance we are dealing with. A few typical figures are shown on page 42.

Thus the whole range of solids vary in *E* by about 200,000 to 1. Even substances which we normally think of as 'rigid' vary by about 1000 to 1, which is still an enormous range. *E* is very low in rubber because rubber is made of long molecular chains which are flexible and in the resting material they are generally much bent, kinked and convoluted, like a heap of bits of string such as one finds in a drawer in the hall at home. When rubber is stretched, the bent chains are straightened and, as one can easily see, the force needed to do so is very much less than that which is needed to stretch an arrangement of strings which were initially straight. Nothing of this kind happens in a normal crystal where one is pulling directly on the interatomic bonds and the only reason for the large variations in Young's modulus is that the chemical bonds themselves vary a great deal in stiffness. So with crystals, although the general shape of the interatomic force curves is

*i.e. by a strain.

Approximate Young's moduli of various substances

	E Pounds per square inch	E MN/m²
Rubber	0.001×10^6 (i.e. 1,000)	7
Unreinforced plastics	0.2×10^6	1,400
Organic molecular crystal, phthalocyanine, a blue pigment	0.2×10^6	1,400
Wood (about)	2.0×10^6	14,000
Concrete	2.5×10^6	17,000
Bone	3.0×10^6	21,000
Magnesium metal	6.0×10^6	42,000
Ordinary glasses	10.0×10^6	70,000
Aluminium	10.5×10^6	73,000
Steel	30.0×10^6	210,000
Aluminium oxide (sapphire)	60.0×10^6	420,000
Diamond	170.0×10^6	1,200,000

Note. Because the interatomic force curve (Figure 3) passes smoothly through the point of zero stress and strain the true E of a material is always the same in compression as it is in tension at all normal strains. If this were not so then the mathematics of elasticity would be even more complicated than they are. In practice, however, materials such as cast iron and cement, which contain quite gross internal cracks, may sometimes show an E which is lower in tension than it is in compression. This is simply because the cracks gape under tension and 'come up solid' under compression.

similar, the slope of the straight part of the curves varies greatly according to the bond energy and other chemical conditions.

The figure for the E of phthalocyanine tells us at once why a great many solid chemical compounds are not candidates for the status of structural materials. Generally speaking we want a structure to be as rigid as possible: bridges and buildings sway quite enough as it is and there are excellent reasons for making other things rigid as well. Any structure made from a material with a stiffness as low as phthalocyanine would be far too floppy. Steel is about the stiffest reasonably cheap material, which is one of the reasons why it is used so widely. As much as anything it is the relatively low stiffness of plastics, even when 'reinforced', which restricts their use for large objects.

Strength

Next to 'heat-proof' I suppose that 'unbreakable' is one of the most useful words in advertising. Although most of us know that advertising is not an entirely objective profession, somehow or other the message sinks in so that one still meets people who really believe that there are unbreakable objects or, if there aren't, then there ought to be. Since there is always some force which will tear the atoms apart in a solid (since the chemical bonds have a finite energy or, in other words, they are only so strong) nothing is unbreakable. You have only to get hold of the thing firmly and pull hard enough and it will break. The only question is 'how soon?' There is however a very large variation between the strengths of various materials.

Lest there be any possible, probable, shadow of doubt, strength is not, repeat not, the same thing as stiffness. Stiffness, Young's modulus or E, is concerned with how stiff, flexible, springy or floppy a material is. Strength is the force or stress needed to break a thing. A biscuit is stiff but weak, steel is stiff and strong, nylon is flexible (low E) and strong, raspberry jelly is flexible (low E) and weak. The two properties together describe a solid about as well as you can reasonably expect two figures to do.

It is easiest to think about strength in terms of tensile strength. This is the stress needed to pull a material asunder by breaking all the bonds between the atoms along the line of fracture. One can perhaps most conveniently think of it as the stress required to break a bar by pulling it along its axis like a rope. A very strong steel may withstand a tensile stress of 450,000 pounds (200 tons) per square inch (3,000 MN/m^2), while ordinary brick or cement may perhaps withstand 600 or 800 p.s.i. or only 4 or 5 MN/m^2.*
The strength of commonly used engineering materials thus varies over a range of about a thousand to one†. The tensile strengths of some common materials are given in the table.

*'Daddy, why can't you make boilers out of cement?'

†For the moment we may be content to say that breaking stress is that stress at which things break. However, let us beware of a trap or incipient muddle. If a bar of 10 square inches cross section breaks under a tensile load of a 100 tons then its breaking load is a 100 tons, but its breaking stress

Some typical tensile strengths in round figures

	p.s.i.	MN/m²
METALS		
Steels		
Steel piano wire (very brittle)	450,000	3,000
High tensile engineering steel	225,000	1,500
Commercial mild steel	60,000	400
Wrought iron		
Traditional	20,000–40,000	140–280
Cast iron		
Traditional	10,000–20,000	70–140
Modern	20,000–40,000	140–280
Other metals		
Aluminium		
cast, pure	10,000	70
alloys	20,000–80,000	140–550
Copper	20,000	140
Brasses	18,000–60,000	120–400
Magnesium alloys	30,000–40,000	200–280
Titanium alloys	100,000–200,000	700–1,400
NON-METALS		
Wood, spruce		
along grain	15,000	100
across grain	500	3
Glass (window or beer-mug)	5,000–25,000	30–170
Good ceramics	5,000–50,000	30–340
Ordinary brick	800	5
Cement and concrete	600	4
Flax	100,000	700
Cotton	50,000	350
Catgut	50,000	350
Silk	50,000	350
Spider's thread	35,000	240
Tendon	15,000	100
Hemp rope	12,000	80
Leather	6,000	40
Bone	20,000	140

is 10 tons per square inch. Engineers refer to the first of these as the strength of the bar and to the second as the strength of the material – fair enough but rather confusing.

When we talk about 'strength' we usually mean tensile strength although materials are more often used in compression than they are in tension. At first sight it is not very easy to see why a material should ever want to break at all in compression. After all, if one is pressing the atoms closer together, why should they come apart? Compressive failure is more complicated than tensile failure, especially as there are several different ways in which a material can run away from a compressive load.

If the material is in the form of a fairly short, squat column, or chock or wedge or something of the sort, then, if the material is at all soft or ductile like mild steel or copper, it will simply squish out sideways, like Plasticine. If the material is brittle, like stone or glass, it will explode sideways (and very dangerous it can be) into dust and splinters. However, if the specimen is long and slender, like a thin rod or panel, then it may fail by 'buckling' such as happens when you lean too hard on a walking stick which bends and ultimately snaps in two. If you put too much weight on a tin can, as by driving a car over it, it will crumple in the same sort of way. This is the kind of failure which is apt to happen to shell structures such as steel ships and metal aeroplanes when they hit things, not to mention the wings of motor cars. For these reasons it is not easy to quote figures in tables for '*the* compressive strength of so and so'. Broadly speaking there isn't one or at least it must be estimated with knowledge and experience. This is one of the reasons why structural engineering isn't particularly easy.

There is no general relationship between the tensile and compressive strengths of various materials and structures, partly because the distinction between a material and a structure is never very clear. For instance, a pile of bricks is strong in compression but has no tensile strength at all. A pile of bricks is undoubtedly a structure and not a material, but then cast iron, cement, plaster and masonry are much stronger in compression than they are in tension and for much the same reason as a pile of bricks: they are full of cracks. Chains and ropes are strong in tension and have no compressive strength because they fold up in compression. They are probably structures not materials. Wood is three or four times as strong in tension as it is in compression because the cell walls fold up in compression, yet wood is thought of as a material not a structure.

Tension and compression structures

For a great many centuries engineers and architects avoided using materials in tension as much as they could. This was not so much because they had no materials strong in tension – wood, for example, is excellent – but rather because of the difficulty of making reliable strong joints to withstand tension. Most of us intuitively feel that a compression structure is safer than a tension one; that a brick tower is safer than an aerial cableway, for instance. In the old days, when tension joints were unavoidable, as they were in ships, they were a perpetual source of trouble. Now that we can make good joints with bolts or rivets, glues or welding, there is no special justification for distrusting a tension structure.

However, in a primitive technology, the problem of a compression joint is very much easier and in the simplest case resolves itself into merely heaping one stone or brick upon another in such a way that the house does not fall down. Dry-walling is a skilled job but not much more so than doing a jigsaw puzzle. As architects became more ambitious and walls higher, it was necessary to arrange firmer, better fitting joints lest the wall slide down with a rumble into a heap of stones. If the stones do not fit each other reasonably well they will roll over each other like a pile of balls and be pushed outward under the superincumbent weight, just as, on a finer scale, Plasticine is pushed outwards. In this way we get magnificently fitted joints between large blocks of stone in ancient buildings. How much of this laborious accuracy was born of engineering necessity and how much of a morbid desire for prestige on the part of men or gods is arguable. Many of these buildings, like a famous car, strike one as being 'a triumph of workmanship over design'.

However high and impressive a wall may be, it remains technologically a very unsophisticated structure because the designer is really only thinking about stresses in one dimension; that is to say vertically. He is always in a difficulty about bridging roofs, doors and other openings. Once he starts thinking imaginatively of stress systems in two and three dimensions, all kinds of possibilities open up, even if he is still restricted to compressive systems. This is why the arch is important. The ordinary simple arch utilizes compression in two directions simultaneously to bridge a gap (Figure 6). This is an apparent

Figure 6. The arch – a two-dimensional compression structure – enables vertical forces to be transmitted laterally around the arch-ring into the abutments. (The wedge-shaped pieces which make up the arch-ring are called 'voussoirs'.)

impossibility which works extremely well. A masonry arch can span 200 feet (60 metres) or so (though 100–200 feet is more common) without much difficulty. This is a very much greater span than any primitive beam or architrave or lintel could bridge. An arch is also durable and there are innumerable Roman arches, such as the aqueducts, in excellent condition today.

The trick of thinking in terms of stresses acting in more than one direction simultaneously is really the key to most advanced architecture and engineering. Once one accepts the two-dimensional concept of the arch or the three-dimensional concept of the dome which is the next logical step, then one can start playing elaborate architectural games. St Sophia, built by Justinian at Constantinople about 530 A.D., is a great dome 107 feet (32 metres) in diameter and made of pumice bricks for lightness, poised upon four great arches which are propped in turn by auxiliary half domes (Figure 7). The result was a nave, completely clear of

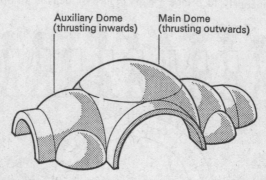

Auxiliary Dome
(thrusting inwards)

Main Dome
(thrusting outwards)

Figure 7. How the outward thrusts of the main dome of St Sophia at Constantinople are supported by means of subsidiary domes and vaults.

pillars, measuring more than 200 feet by 100 feet (60 metres × 30 metres) and about 240 feet (72 metres) high, a clear roofed area greater, probably, than any achieved until the advent of the modern railway station which is roofed with steel trusses. The shapes of Byzantine architecture are usually simple but the Gothic architects ran riot in aisles, fan-vaulting and clerestories. All this, though rather expensive, is great technical and artistic fun as long as you know what you are doing. The essential thing about a masonry structure is that it must be a compression structure everywhere, because masonry is incapable of resisting tensions since the stones will come apart at the joints.

In the three-dimensional labyrinth of a cathedral roof, where thrust chases thrust in Gothic disregard of mathematics, strange things were apt to happen. Tensions crept in, like devils among the gargoyles. In one of the greatest of the Gothic cathedrals, Beauvais (1247), the tower fell once and the roof fell twice. Contemporary architects knew what was wrong in a qualitative sort of way and they propped their structures up with a maze of flying buttresses (Figure 8), just as St Sophia is set about, in a more rational and successful way, with auxiliary domes which thrust inwards and maintain a state of compression in the critical regions. Sometimes the Gothic architects overdid the business of

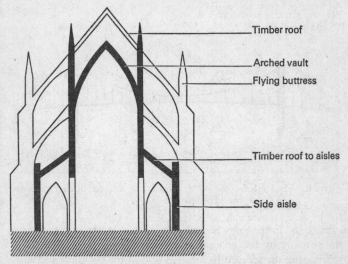

Timber roof

Arched vault

Flying buttress

Timber roof to aisles

Side aisle

Figure 8. In a 'Gothic' structure the outward thrust of the roof is taken mainly by buttresses.

inward thrusting and had to strut their naves internally to prevent the building collapsing inwards. This strutting was sometimes done by inserting inverted arches, as at Wells Cathedral (Plate I) which, whatever may be thought of it aesthetically, is a mess structurally. It is not surprising that the roofs of churches continued to fall upon the heads of their congregations with fair regularity throughout the ages of faith.

A masonry structure is kept together by gravity; that is, if it is properly designed, the weight of the stones keeps everything safely in compression. If necessary one can pile on pinnacles and towers to get more weight in the right place. Once we start dealing with tension, however, or mixed tension and compression structures, we have to accept that the tensions and compressions must balance out, taking into account the weights of the various parts, of course. In a suspension bridge the cables are maintained in tension by a corresponding horizontal compression in the ground beneath the bridge (Figure 9). In a tent the tension in the

Suspension cables in tension
Towers in compression

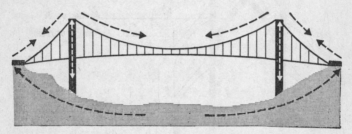

Figure 9.   The tension in the cables is reacted by a corresponding compression in the ground beneath.

canvas and in the guys is reacted vertically by the tent pole and horizontally by the ground on which the tent is pitched. In a sailing ship the tension in the sails and in the standing and running rigging is reacted by compressions in the masts and spars. In an animal the bones, and especially the backbone, are chiefly compressive members reacting, not only the weight of the animal, but also the tensions in the muscles and tendons. I raise my arm by shortening a muscle, that is to say by pulling on it, and this puts the bone into compression – which is what bones are usually designed to take. Getting one's leg into bending, which involves tensions, is the easiest way to break it.

It is a great convenience and a great source of safety to be able to pass, as it were, from compression into tension and back again, either deliberately or by accident. In architecture this is one of the arguments for reinforced concrete and for steel-framed buildings, both materials being strong in tension and compression. It is also one of the reasons why iron and steel are such a godsend to engineers, putting a cloak over their ignorances and uncertainties.

Boilers, for instance, are tension structures which may occasionally get into compression (if you let the fire out you can drive an
engine under the negative pressure, that is the vacuum in the
boiler) without anything very dangerous happening.

The compressive stresses in a submarine can lead to some
rather interesting and unexpected effects of strains, which have to
be thought about and guarded against. When a submarine is on
the surface she floats, like any other ship, because the weight of
the submarine is less than the total weight of water which would
be displaced if the vessel were totally submerged. If, for any
reason, the hull sinks a little in the water, a greater volume is
immersed and the extra buoyancy pushes the vessel up again.
When a submarine dives, she fills her ballast tanks with water
until her total weight just about equals her submerged displacement and so she has no reserve of buoyancy. In this condition she
can dive and manoeuvre under water in much the same way as an
airship does in the air. However, as the submarine dives deeper,
the water pressure increases and the hull is put under more and
more compression. Because the air inside the hull is not under
pressure the steel in the hull can only resist this compression by
contracting. So the volume of the hull, and thus its displacement,
is reduced, although the weight of the submarine and her ballast
water is, naturally, not changed. There is therefore a tendency for
the submarine to sink further, or to become apparently heavier,
the deeper she goes – and in certain circumstances this can be
dangerous.

At the safe limit of diving depth the compressive strain in the
hull plates might be about 0·7 per cent. Since this strain occurs
in all three directions the hull may shrink by about 2 per cent in
volume. As water is only very slightly compressed, this may
represent a loss of about 20 tons or so of buoyancy for a 1,000-
ton submarine.* If this weight cannot be counteracted by blowing
the ballast tanks or working the hydrofoils, the submarine will
sink deeper and deeper until she is crushed by the water pressure

* Or tonnes – near enough. The total weight of ballast water needed in
order to enable such a submarine to dive might be around 300 tons. The
elastic contraction of the hull may therefore call for quite substantial proportionate adjustments in ballast.

in the depths of the ocean. This is one of the difficulties about making submarines out of reinforced plastics, such as fibre-glass, which are otherwise rather attractive, but which have low Young's moduli. It is nonsense to think, as used sometimes to be said, that sinking submarines and wrecks will float somewhere short of the bottom of the sea. Even if the pressure hulls and compartments which contain air do not actually burst inwards, which must usually happen, they will progressively contract and lose buoyancy and so the wreck sinks faster and faster.

Balloons, pneumatic tyres and the like are a special case of a tension structure where the tension in the skin is reacted by the pressure of the gas or liquid inside. In this way Dracones (large bag-like barges for conveying liquids) and pneumatic boats are usually very light and efficient structures. The air-supported roof – a building held up entirely by internal pressure – reverses architectural tradition in that everything except the air inside is in tension. Since only a very small air pressure is needed, a modest electric blower keeps everything taut and even supports any reasonable snow load for less expense than the capital charges of a conventional building. Plants and animals use the osmotic pressure of their internal fluids in a similar way.

Beams and bending

It is quite easy to see how tension and compression structures work but it is not at all self-evident how the tensions and compressions which we have been discussing really support a load on a beam. This is a pity, since beams of one kind or another (Figure 10) make up a high proportion of everyday structures. The ordinary floor-board is as good an example as any other of a simple beam. As we said earlier it is the function of a floor-board to press upwards on the soles of our feet with a thrust equal to our own weight. Yet it must perform this function even when we are standing in the middle of the floor and the walls which eventually support the plank are remote. Exactly how does this thrust get from the wall to our feet and vice versa?

The answer to this question is known as 'beam theory' to engineers and is more or less the backbone of engineering. Unfortunately it is less of a backbone than a *pons asinorum* to engineering students. Most students merely learn the formulae of beam theory off by heart and regurgitate them at examination times; understanding only comes much later when they have to struggle

Figure 10. Simply supported beam.

with designing something. We shall therefore leave out all that stuff about integrating the shearing force diagram and try to tackle the problem by the light of nature. Once beam theory is understood the world becomes a more lucid and an altogether better place, so take courage.

To understand about beams it is perhaps easiest to return to the idea that there is no very clear distinction between a material and a structure. Large beams are often fabricated, Meccano fashion, from many small tension and compression rods, as anyone can see who looks at a railway bridge. Yet the means by which the load is transmitted in such a lattice beam or girder is not different in kind from the means by which it is transmitted in a solid beam, even so humble a one as a plank or floor-board. In the lattice structure we can generally see the individual members which resist all the pushes and pulls – all of them, since of course no load can get across the empty spaces between the lattice-work. In a solid beam we have to consider the lattice members as diffused throughout the beam, but the stresses are working in the same way.

We might as well start with a cantilever, a beam one end of which is built into a wall or otherwise fixed to a firm base (what engineers call *encastré*) while it is loaded in some way on the

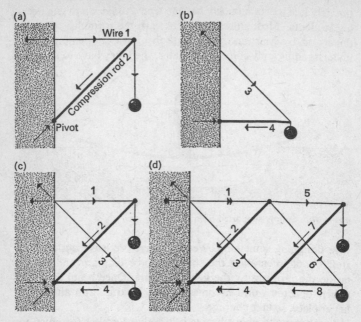

Figure 11.   Beam theory – a beam may be considered as made up from a number of separate panels.

projecting part. Galileo's picture (Plate 2) of a cantilever will serve as well as any other although Galileo, rather excusably, got his cantilever sums wrong. Let us, however, build up a cantilever entirely by tension in wires and rods.

Consider the simple crane-like structure in Figure 11(*a*). A compression rod (2) is pivoted against a firm wall and is supported by a wire or tension member (1), so that it can carry at its outer end a load, *W*, say. Now it is clear that the push against gravity which is actually supporting the load *W* is generated from the compression in the sloping rod (2). The tension in the wire (1) acts horizontally and only prevents the compression rod (2) from rotating and falling down.

Now we might equally support the weight *W* from another triangular structure such as (*b*) in which the compression rod (4)

(e)

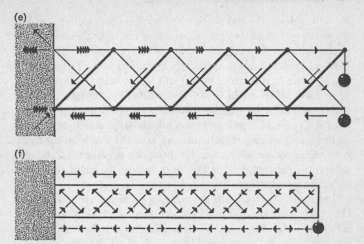

(f)

But the stress system in a solid cantilever is not very different from that in a lattice truss.

(g)

And, of course, in order to support the weight, the material must be strained and so the cantilever will droop.

was horizontal and was kept from falling down by the diagonal tension wire (3). In this case the upthrust to support the weight W comes from the slanting wire (3) and all the horizontal compression rod does is to prevent the wire from collapsing inwards onto the wall.

These two structures are each as good as the other and we might combine the two to support $2W$, as in (c). Clearly the weight $2W$ is directly supported by the two slanting members (2) and (3), one in tension and the other in compression. The horizontal members (1) and (4) pull and push on the wall to prevent the whole structure swinging downwards but their thrusts do not directly sustain the weight.

We can now repeat this structure by duplicating (c) so as to get (d). Here we have a lattice girder with two panels. In this case the same load $2W$ is again actually supported by the tensions and compressions in the slanting members (2), (3), (6), and (7) while (1), (5), (4), and (8) pull and push horizontally, and though they do not directly support the load, they keep the whole girder from collapsing; indeed every member performs an essential, though different, function and the failure of any of the eight members would be catastrophic.

Notice the way the loads are building up in our simple girder. The outer panel in (d) is in all respects similar to the single panel in (c). Consider, however, the inner panel in (d); that is the one next the wall. A little consideration will show that the tension in the wire (1) is greater that in (5) and in the same way the compression in (4) is greater that in (8). This is because the diagonal or 'shear' members are feeding load in progressively towards the root of the cantilever. In the shear or slanting members, however, the loads are the same in each panel, however long the girder may be.

We can go on and build up a long girder of many panels like (e) and here again, if we look at the lattice intelligently, it is obvious that the load in all the diagonal shear members is constant along the length of the girder, however many panels there may be. On the other hand, the tensions and compressions in the top and bottom horizontal members (called the booms or flanges of the girder) are building up and increasing as we move from the loaded tip to the built-in root of the girder, in fact in proportion to its length. For this reason a cantilever will usually break in its most highly stressed members – the horizontal members which are up against the wall – unless we have gone to the trouble of making the thickness of each part proportional to the load which it has to carry. In such a case the lattice may break anywhere, which is an ideal state of affairs and the aim of most stress calculations. When this occurs all the material is equally stressed (like the one-horse shay) and the material is used in the most efficient way. Hence the least quantity of material can be used and the lightest structure will result.

If we now convert our lattice or Meccano girder to a simple continuous beam we shall get a stress system like (f). The middle of the beam is mostly occupied with resisting shear, which turns out to be another name for tensions and compressions at forty-five degrees to the axis. These shear stresses are of constant magnitude for the whole length of the beam. The material near the top and bottom surfaces is concerned with resisting the tensions and compressions which the shear stresses have generated. These horizontal stresses near the surfaces build up rapidly and in the worst place are usually far greater than the shears. They are the stresses which are most liable to break structures and kill people. Stressing is not a dry academic exercise of interest only to experts but an affair affecting the safety and pockets of most of us.

If all this stress-chasing seems confusing the best thing to do is to make a model out of Meccano or drinking straws joined with ordinary pins. If one makes a lattice model in this way it is quite easy to understand what it is in a cantilever which actually keeps the load from falling down. Of course, in all this rather complicated pattern of stresses, each stress is only gained at the expense of a proportionate strain and so the cantilever does not stick out rigidly but inevitably droops (g) to a greater or less extent.

Cantilevers are common enough in engineering but ordinary beams, particularly those that engineers call 'simply supported', are commoner still (Figure 10). This is the sort of thing you get when you put a plank across any simple gap such as a stream. How is this related to the cantilever? The answer is really quite obvious from Figure 12. The simply supported beam is really two cantilevers turned back to back and upside down. While the biggest stresses in the cantilever are near the root, those in a simply supported beam are in the middle and so such a beam will generally break in the middle.

We can now see that the reason why we don't fall through the floor is that the floor boards and joists produce tensions and compressions at forty-five degrees to the surface of the floor and

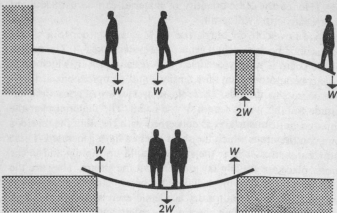

Figure 12. A simply supported beam may be considered as two cantilevers back to back and upside down.

these stresses, elaborately zig-zagging all the way from my shoes to the skirting-board, provide the upward sustenance which I need. As well as these shears, and much larger in magnitude, there are tensions and compressions horizontally near the top and bottom faces of the floor boards. If these horizontal stresses become too large, either because I am too heavy or the floor boards are too thin we shall first get alarming deflection in the floor and finally it will break.

The simplest experiment will show that the stresses and deflections induced by bending are, other things being equal, much more severe than those caused by direct tension and compression. If we take a piece of fairly thin wooden plank or rod in our hands, it is quite impossible to break it in tension by pulling on it by hand and the deflections we can cause by hand tension or compression are far too small to see by eye. If we bend the rod, however, we can nearly always produce quite a big and obvious deflection and in many cases it is quite easy to break it. For these sorts of reasons, although beams are extremely convenient, we nearly always have to be careful that they are strong enough and don't produce excessive deflections. The strengths and deflections of a given beam can be calculated by anyone with a knowledge of elementary algebra from the standard formulae which are given in the appendix at the end of this book.

As we have said, all this is not particularly easy to understand but it is really no more difficult than, say, French verbs, and it can be comprehended by quite a moderate intellectual effort. Once this is done, a great deal of engineering becomes much clearer. The truth is that many professional engineers use very little more than elementary beam theory when they design quite ambitious structures. As we shall see this is apt to be dangerous because beam theory in itself, though extremely useful, does not really tell us all we need to know about the strength of a sophisticated structure. However, it is very widely used as a guide to the strength of all kinds of things from crankshafts to ships.

The deliberate and confident use of large beams in engineering is not much more than a century old. Telford (1757–1834) – the 'Colossus of roads', or 'Pontifex Maximus', as Southey called him – probably built more bridges than anyone else in history. He used masonry or cast-iron arches in compression and for the longer spans he pioneered the suspension bridge, using wrought-iron tension chains, notably in the Menai road bridge (1819). Telford hardly ever used large beams. This was partly because a suitable material, such as wrought-iron plates, was not easily available and partly because of the lack of a trustworthy beam theory. An interesting sidelight on the status of strength calculations in Telford's time is that the shape of the chain catenaries for the Menai bridge was determined not by calculation but by setting up a large model across a dry valley.

Working thirty years later, Robert Stephenson (1803–59) had large wrought-iron boiler plates available and he also had the courage of his calculations. He had the brilliant idea* of making a hollow box-like beam of iron plates and running the trains inside it. The idea found its best-known expression in the Menai railway bridge which was opened in 1850, almost alongside Telford's bridge. Stephenson's beams, which weighed 1,500 tons each, were built beside the Straits and were floated into position

* Actually much of the credit must go to Stephenson's designer, William Fairbairn.

between the towers on rafts across a swirling tide. They were raised rather over a hundred feet up the towers by successive lifts with primitive hydraulic jacks. All this was not done without both apprehension and adventure; they were giants on the earth in those days.

At one time, when Stephenson's faith weakened, it was proposed to add suspension chains to help sustain the tubes, but this proved quite unnecessary. Both bridges stood side by side until recently as elegant demonstrations of tension and bending upon the grand scale.* Telford's suspension bridge was lacking in stiffness at first and the bridge swayed alarmingly in the gales which blew down the Straits. There is an account of how one winter night the oscillations were such that the horses of the mail-coach could not keep their feet on the bridge and were thrown down in a dangerous tangle of hooves and harness so that the traces had to be cut by lantern light before the mess could be sorted out. After this the bridge was stiffened and now carries modern traffic.

The general lack of stiffness of suspension bridges made them unsuitable for railways, since the trains might have been rolled off the rails. This was why Stephenson and I. K. Brunel (1806–59) developed beam-like bridges for long spans. Though the Menai tubular bridge is splendidly stiff and has never given any trouble, equivalent modern beam bridges are generally lattice structures because lattices are easier to paint and 'To keep the Menai bridge from rust by boiling it in wine' was impracticable.

A ship is a long tube closed at both ends which happens to be afloat but is not otherwise structurally very different from Stephenson's Menai bridge. The support which the water gives to the hull does not necessarily coincide with the weights of engines, cargo and fuel which are put into the ship and so there is a tendency for the hull to bend. It ought to be impossible to break a ship, floating alongside a quay, by careless and uneven loading of the holds and tanks, but this has happened often enough and will probably happen again. In dry-dock ships are supported with care upon keel-blocks arranged to give even support but there is not much even support at sea where a ship may be picked up by

* Unfortunately the Britannia bridge was destroyed by fire in 1970 and has been replaced by a steel arch bridge.

rude waves at each end, leaving her heavy middle unsustained, or else exposing a naked forefoot and propeller at the same moment.

As ships tended to get longer and more lightly built, the Admiralty decided to make some practical experiments on the strength of ships. In 1903 a destroyer, H.M.S. *Wolf*, was specially prepared for the purpose. The ship was put into dry-dock and the water was pumped out while she was supported, in succession, amidships and at the ends. The stresses in various parts of the hull were measured with strain-gauges, which are sensitive means of measuring changes of length, and therefore of strain, in a material. The ship was then taken to sea to look for bad weather. It does not require very much imagination to visualize the observers, struggling with seasickness and with the old-fashioned temperamental strain-gauges, wedged into Plutonic compartments in the bottom of the ship, which was put through a sea which was described in the official report as 'rough and especially steep with much force and vigour'. Her captain seems to have given the *Wolf* as bad a time as he could manage but, whatever they did, no stress greater than about 12,000 p.s.i. or 80 MN/m² could be found in the ship's hull.

As the tensile strength of the steel used in ships was about 60,000 p.s.i. or 400 MN/m², and no stress anywhere near this figure could be measured, either at sea or during the bending trials in dry-dock, not only the Admiralty Constructors but Naval Architects in general concluded that the methods of calculating the strength of ships by simple beam theory, which had become standardized, were satisfactory and ensured an ample margin of safety. Sometimes nobody is quite as blind as the expert.

Ships continued to break from time to time. A 300-foot (90 metres) ore-carrying steamer, for instance, broke in two and sank in a storm on one of the Great Lakes of America. The maximum calculated stress under the probable conditions was not more than a third of the breaking stress of the ship's material. Even when major disasters did not actually happen, cracks appeared around hatchways and other openings in the hull and decks.* These

* In 1966 the honorary vice-president of the Royal Institution of Naval Architects, Mr J. M. Murray, announced 'Since 1950 only 26 ships have broken in two on the high seas'. Of a considerable number of ships which

openings are of course the key to the problem. Stephenson's tubular bridge was eminently satisfactory because it is a continuous shell with no holes in it except the rivet holes. Ships have hatchways and all sorts of other openings. Naval Architects are not especially stupid and they made due allowance for the material which was cut away at the openings, increasing the calculated stresses around the holes *pro rata*. Professor Inglis, in a famous paper in 1913, showed however that '*pro rata*' was not good enough and he introduced the concept of 'stress-concentration' which, as we shall see (Chapter 4), is of vital importance both in calculating the strength of structures and in understanding materials.

What Inglis was saying was that if we remove, say, a third of the cross-section of a member by cutting a hole in it then the stress at the edge of the hole is not $\frac{3}{2}$ (or 1·5) of the average but it may, locally, be many times as high. The amount by which the stress is raised above the average by the hole – the stress-concentration factor – depends both upon the shape of the hole and upon the material, being worst for sharp re-entrants and for brittle materials. This conclusion, which Inglis arrived at by mathematical analysis, was regarded with the usual lack of respect by that curiously impractical tribe who call themselves 'practical men'. This was largely because mild steel is, of all materials, perhaps the least susceptible to the effects of stress concentrations though it is by no means impervious (Plate 3). It is significant that, in the *Wolf* experiments, none of the strain gauges seems to have been put close to the edge of any important opening in the hull.

Note: problems of walls, arches, beams and so on are dealt with more fully in the author's *Structures*, Penguin Books, 1978.

---

had come to no harm at sea but which were examined in dock, 20 per cent were found to have cracks in the main hull girder.

Chapter 3 Cohesion

or how strong ought materials to be?

> '*Again, the things that we see to be hard and dense must needs consist of particles more mutually hooked and must be deeply held compact by branch-like elements. In this class, for example, stands adamantine rock, accustomed to laugh blows to scorn, and stalwart flint, and the hard strength of iron, and the copper bolts that scream as they resist their rooves.*'*
>
> Lucretius, *De rerum natura*.†

Before one can start arguing about how strong materials ought to be one should be able to measure how strong they actually are. Although nowadays a certain amount of mechanical testing is done for what might be called academic reasons, by far the most of it is done for strictly practical ends and in fact a thorough knowledge of the actual strength of its materials is, like drains and income tax, one of the things which no advanced civilization can do without.

There are generally two pragmatic reasons for knowing the strength of a material. The first and the most obvious is to have a figure to put into one's calculations on the strength of structures. However since proper scientific strength analysis is a recent affair, much the older and the commoner is that of maintaining the quality of materials. In other words, is this batch as good as the last? A variant of this is, can I use this as a substitute for that?

Of course anything as scientific as a mechanical test has not always found favour with traditional craftsmen or indeed with

---

*The word is 'claustrum' which means a closing or something closed. Jackson translates it 'staples' which may be correct if Lucretius is talking of door bolts. If however he means structural bolts then the word must mean 'rooves' which are the washers put under the clenched or riveted ends of copper bolts.

† Book II, 444, translated by Thomas Jackson.

business men.* The procedure described in Weston Martyr's (1885–1966) delightful book *The Southseaman*† (which is about wooden shipbuilding in Nova Scotia in the nineteen twenties) was probably much commoner.

Before any plank was put into place, MacAlpine and Tom and anyone else who happened to be about held a consultation over it. First they examined it very carefully, and then they bent it, tapped it, listened to it, and, as I live by bread, I swear that once, at least, I saw MacAlpine tasting it. At any rate he applied his tongue to the wood, and then went through all the motions of an expert tea-taster – even to that final feat of expectorating through the clenched teeth with precision and gusto.

The first published tensile tests seem to have been done by the French philosopher and musician Marin Mersenne (1588–1648) who was interested in the strength of the wires used in musical instruments. In 1636 Mersenne made a series of tests on wires of different materials but it is doubtful if any use was made of the information.

As far as I know, the first actual record of an objective mechanical test, which had results of practical consequence, occurs in Pepys' diary for 4 June 1662.

Povey and Sir W. Batten and I by water to Woolwich; and there saw an experiment made of Sir R. Ford's Holland's yarn (about which we have lately made so much stir; and I have much concerned myself of our rope-maker, Mr Hughes who represented it so bad) and we found it to be very bad, and broke sooner than, upon a fair triall, five threads of that against four of Riga yarne; also that some of it had old stuffe that had been tarred, covered over with new hempe, which is such a cheat as has not been heard of.

The Woolwich people may have broken these ropes in direct tension by hanging weights on them, having tied some sort of scale pan to one end and the other end to an overhead beam. On the whole, however, it is more likely that they used a comparative

---

*According to *The History of the British Iron and Steel Industry* (by J. C. Carr and W. Taplin, Harvard University Press, 1962) the leading British ironmaster of the 1870s used to say 'I know nothing about tests; if they want my brand they can have it; if they don't they can go elsewhere.'

† *The Southseaman*, J. Weston Martyr, Blackwoods, 1928.

test, tying the two competing ropes end to end, in series, and breaking them by means of a capstan. The number of strands in each rope would then be adjusted until there was an equal chance of failure.

Ropes and wires are fairly simple to test since it is easy to grip the ends by winding them round the barrel of a winch or capstan. Rigid solids are much harder to get hold of in tension and so for a long time such testing as was done was confined to compression and bending. Testing machines now exist which have vice-like grips, called 'friction grips', so that one can take an ordinary bar of metal, cut off a short length, and break it in tension. In practice however, this is generally an unsatisfactory arrangement since the grips damage the metal and cause premature failure at the ends so that the result is unreliable. It is usually better to cause an hour-glass or wasp-waisted specimen to be made as this can be arranged to break in the middle where it is thinnest. Even so, the design and making of satisfactory test-pieces calls for a modest degree of skill and experience as the best shape will be different for each kind of material.

With regard to the actual mechanics of testing, it is of course possible to apply the load to each specimen directly, by means of weights. However, since the breaking loads for convenient sized test pieces (say $\frac{1}{4}$ inch thick) are typically between about one and ten tons (a motor car weighs about a ton) and since most testing is done by girls, it is usual to apply the load mechanically or hydraulically and there are a large number of more or less automatic testing machines on the market. All that the operator has to do is to insert the specimen, watch the machine break it, then divide the recorded breaking load by the area of the cross-section at fracture, which is easily measured. The result is the breaking stress.

Of course this figure tells one nothing at all about why the material has the strength which it has and whether it ought to be stronger. On the other hand, in practice, the strength of any one individual engineering material tends to be constant. There therefore grew up a tendency to ignore the whys and wherefores and to regard the tensile and other strengths as innate properties with which the material happened to have been endowed by

Providence in a rather arbitrary way. Metallurgists knew that this or that ingredient or heat treatment would strengthen or weaken an alloy but this knowledge was empirical and the effects were not susceptible to an obvious rational explanation.

Engineers like their materials to be consistent and are not too deeply interested in reasons, so they encouraged the idea that each material has a characteristic strength which could be determined accurately, once for all, if only one did enough tests. Materials laboratories of a generation ago centred upon magnificent collections of large testing machines. We filled a great many notebooks with testing data but learnt very little about the strength of materials.

Indeed it is difficult to exaggerate the impenetrability of the mystery which for centuries hung over the problem of the strength and fracture of solids. Lucretius (95–55 B.C.) set forth at great length the theory of the atomic nature of matter which had been propounded earlier by Democritus (460–370 B.C.). Though the theory was many years ahead of its time it was almost wholly guesswork and rested on no satisfactory contemporary experimental evidence. However Lucretius recognized the problem of cohesion and suggested that the atoms of strong materials were provided with hooks with which to grip each other. In the middle of the nineteenth century Faraday (1791–1867), one of the wisest of men, could do no better than to say that the strength of solids was due to the cohesion between their fine particles and that the subject was a very interesting one. Though both these statements were true they were not a great advance on Lucretius.

Chapter 2 contains a list of the practical strengths of various materials. Like the values for Young's modulus or stiffness, the figures vary a great deal between different substances, but then, so do the strengths of the chemical bonds within them and one might expect the engineering strengths to be proportional to the strengths of the chemical bonds. This is one of the differences between strength and stiffness. One can relate the Young's modulus, E, for a material in bulk to the fine-scale stiffness of its chemical bonds with considerable accuracy. Generally speaking this is not true of strength. The iron to iron bond in steel is not especially strong, it is easily broken chemically when iron rusts.

Rust, iron oxide, is weak mechanically although its chemical bonds are strong. Again, magnesium metal is stronger than magnesium oxide, magnesia, though the energy difference in the bonds is dramatically shown by burning magnesium ribbon in oxygen. Any attempt to relate chemical to mechanical strength works only in a vague and irregular way. About all one can say is that while it is only too easy to make a weak material (or indeed a material of no strength at all) from strong chemical bonds, it is not possible to make very strong materials from weak bonds.

The plastics and polymers which came into use between the wars were, or were claimed to be, the first man-made strong materials to come out of chemical laboratories and they rather went to the heads of the chemists, who supposed, not unnaturally, that these polymers were strong because they had put them together with strong chemical bonds. When the last war broke out, a very able young academic chemist came to work with me. He set to work straight away to make a stronger plastic. He explained to me that it must be stronger because it contained stronger bonds and more of them than any previous material. Since he really was a very competent chemist I expect it did. At any rate it took a long time to synthesize. When it was ready we removed this war-winning product from the mould with excitement. It was about as strong as stale hard cheese.

Griffith and energy

We must now go back to about 1920 when the whole subject could be described as pretty well bogged down. At this time A. A. Griffith (1893–1963) was a young man working at the Royal Aircraft Establishment at Farnborough. He had ideas which cut through the mass of tradition and very dull detail which hung around materials work everywhere but unfortunately nobody took them very seriously. Griffith asked in effect 'Why are there large variations between the strengths of different solids? Why don't all solids have the same strength? Why do they have any strength at all? Why aren't they much stronger? How strong "ought" they to be anyway?' Until fairly recently these questions were regarded as unfathomable or unimportant or just silly.

We now understand in a general way how strong any particular solid ought to be and why it falls short of that strength in practice. Furthermore we know more or less what to do to increase its strength. Much of this success is due directly and indirectly to Griffith. In what follows I have shortened and transposed Griffith's arguments.

To calculate how strong a material ought to be we need to make use of the concept of energy. Energy is officially defined as 'capacity for doing work' and it has the dimensions of force multiplied by distance. Thus if I raise a two-pound weight through a height of five feet I have increased its potential energy by ten foot pounds.* This energy (which comes from my dinner which comes ultimately from the sun and so on) can be transformed into any of the many alternative forms of energy but it cannot be destroyed. Potential energy is one convenient way of parking energy until it is wanted and this energy can be followed through its various subsequent transformations by a sort of accounting procedure which can be very revealing.

The stored or potential energy in a raised weight can be used, for instance, to drive the mechanism of a grandfather clock though in most clocks a spring is usually more convenient, if only because it stores the same amount of energy which ever way up it is.† The strain energy in a stretched material is very like the potential energy which is in a raised weight, except, of course, that the stress is changing as the material is strained whereas the

*The S.I. unit of energy is the Joule which is the work done when 1 Newton acts through 1 metre. 1 Joule (1J) = 10^7 ergs = 0·74 foot–pound = 0·23 calories.

†A clock spring is really a beam – a flat strip of steel wound up into a spiral, like a tape measure, to save space. When the clock is wound, more turns are put into the spiral and, since the total length of the spring remains the same, any given short length is more sharply bent. Being a beam, every small element of the material of the spring is either in tension or compression (Chapter 2).

Incidentally, because a spring produces less energy per turn as it runs down, the early clockmakers had to invent a device called a 'fusee', a sort of conical spool on which was wound a driving chain, to keep the driving force constant. This was why they generally preferred the grandfather clock for time-keeping since an inch of fall of the weight is good for as much energy at the bottom as at the top of the case.

weight of a weight is constant as it is raised to any normal height.

Because of Hooke's law, when a material is strained the stress in it varies from nothing at the beginning of the operation up to a maximum at the final strain. For this reason the strain energy in a material is:

$\frac{1}{2}$ stress × strain   per unit volume

That strain energy is more than a triviality was demonstrated by the bowmen at Agincourt and, incidentally, one is well advised to keep out of the way of a stretched hawser such as is used for checking a ship. The kinetic energy, or energy of motion, of the ship has been exchanged for strain energy in the rope. There is a lot of energy, and, if the rope breaks strain energy is reconverted to energy of motion in the rope and somebody may get killed.

All stressed solids thus contain strain energy and this strain energy can be converted by one means or another into any of the other forms of energy. Most commonly a relaxed stress simply reverts to heat but children have discovered that it is possible to convert the strain energy of catapult elastic into the fracture of, say, glass. Whether or not something of the sort put into Griffith's head the idea of fracture as an energy process, I have no idea.

When a brittle material breaks, two new surfaces are created at the point of fracture which were not there before fracture, and Griffith's very brilliant idea was to relate the surface energy of the fracture surfaces to the strain energy in the material before it broke. Energy has many forms – heat, electrical energy, mechanical energy, strain energy and so on – but it is not immediately clear that the surface of a solid has energy, merely by virtue of its existence as a surface.

From watching raindrops, bubbles and insects walking on ponds it is obvious that water and other liquids have a surface tension. This tension is a perfectly real physical force which is quite easily measured. Consequently, when the surface of a liquid is extended, as by inflating a soap bubble, work is done against this tension and energy is stored in the new surface. In the accountancy of energy, surface energy counts in the balance just as much as any other kind of energy. When an insect alights on water, the surface is dimpled by its legs and thus extended and so the surface energy is increased. The insect sinks until the increase of surface energy just balances the decrease in its potential energy when it sinks no further and is, presumably, happy. Liquids tend, if they can, to minimize their surface energy. For instance a thin stream of liquid, from a tap which is being turned off, will reach a diameter at which it pays it to break up into separate drops simply because these have less aggregate area then the cylindrical stream.

When a liquid freezes, the molecular character of its surface is not too greatly changed and the energy of the surface remains much the same although the surface tension is no longer able to change the shape of small particles by rounding them off into drops. With a number of solids the interatomic forces are stronger and stiffer than they are in common liquids and so the surface energies are higher, often ten or twenty times the values for ordinary liquids.* The reason why we do not notice surface tensions in solids is not that the surface tensions are weak but rather that solids are too rigid to be visibly distorted by them.

Just as we could, perhaps, calculate the weight of the largest insect which could walk on a given liquid, so we can use these concepts to calculate how strong we ought to expect materials to be. As calculations go, this one turns out to be surprisingly simple, once somebody has had the original idea.

What we want to do is to calculate the stress which will just separate two adjacent layers of atoms inside the material. At this stage we need not worry too much whether the material is glassy or crystalline and all we really need to know about the solid is the Young's modulus and the surface energy. The two layers of atoms are initially x metres apart and so the strain energy per square metre for a stress s causing a strain e will be:

$$\tfrac{1}{2} \text{ stress} \times \text{strain} \times \text{volume} = \tfrac{1}{2} \, s.e.x.$$

*The surface energy of water is about 0·077 Joules per square metre. Structural solids have usually energies around 1·0 J/m². Surface energy of diamond = 5·14 J/m². N.B. The surface *energy* in Joules per square metre is *numerically* equal to the surface *tension* in Newtons per metre.

But Hooke's law says:

$$E - \frac{s}{e} \text{ so } e = \frac{s}{E}$$

So, putting in $\frac{s}{E}$ for e:

$$\text{Strain energy per square metre} = \frac{s^2.x}{2E}$$

If G is the surface energy of the solid per square metre, then the total surface energy of the two new fracture surfaces would be $2G$ per sq. metre.

We now suppose that, at our theoretical strength, the whole of the strain energy between any two layers of atoms is potentially convertible to surface energy, then:

$$\frac{s^2.x}{2E} = 2G$$

so:

$$s = 2\sqrt{\frac{G.E}{x}}$$

Actually, this is a bit optimistic because we have assumed that the material will go on obeying Hooke's law right up to failure. As we saw in the last chapter, Hooke's law is really only true for small strains and at large strains the interatomic force curve bends over so that the strain energy is less than we have calculated, very roughly about half. We can allow for this effect by dropping the two from the strength equation which we have just derived, bearing in mind that we are in no position to quibble about exact values. Thus a reasonable expectation for the strength of a material would be:

$$s - \sqrt{\frac{G.E}{x}}$$

which could hardly be much simpler.

For steel some typical values in S.I. units would be:

Surface energy $G = 1J$ per square metre
Young's modulus $E = 2 \times 10^{11}$ Newtons per square metre
(note *not* Meganewtons)
Distance between atoms $x = 2$ Ångström units
$= 2 \times 10^{-10}$ metre

Putting in these values gives us a strength of about 3×10^4 MN/m² or about five million pounds per square inch. Say ($E/6$). This is rather over 2000 tons per square inch. The strength of ordinary commercial steels is usually about 60,000 p.s.i. or 400 MN/m² while very strong wires may reach about 400,000 p.s.i. or 3,000 MN/m².

Since the values for E and G vary, of course, for each solid the values we get for the theoretical strengths will vary too. The only thing they have in common is that they are all very much above any strength normally realized in ordinary experiments. In fact steel is exceptional in sometimes reaching strengths as high as a tenth of its calculated strength; the great majority of common solids can show only a hundredth or a thousandth of what theory indicates.

As a matter of fact, thirty or forty years ago, nobody actually and openly disbelieved this calculation. If they had, they would have had to provide an alternative explanation of where the surface energy of a newly broken surface came from, but somehow nobody took it very seriously. There was a discrepancy somewhere and perhaps the less said about it the better.

If we confine our calculation simply to strength as such, we get a different figure for the theoretical strength of each material. However, we can nearly as easily do the sum for the theoretical elastic breaking strain, and if we do this, we are apt to find that the answer we get is very roughly the same for any solid, almost irrespective of its chemical entity. Generally speaking, this strain is something like 10 or 20 per cent.* If this is more or less

* Discerning minds will deduce that the surface energy of a solid must be roughly proportional to its Young's modulus – and so it is. G is more or less equal to $Ex/20$. This arises because it is the same bonds which give rise both to the Young's modulus and to the surface tension.

true, then the strength of any solid should lie between $E/10$ and $E/5$. Hence, although we cannot say that every material ought to have the same strength, we can say that, very approximately, all materials ought to have the same elastic breaking strain. In everyday practice it is palpable that not only do materials not have the same breaking strain but also that the calculated strengths are, without exception, far above any commonly realized practical strength.

Griffith set out to find some physical theory which would bridge this gap between theory and practice. I never knew Griffith himself but Sir Ben Lockspeiser, who acted as Griffith's assistant at this time, told me something about the circumstances under which the work was done. In those days research workers were expected to earn their money by being practical, and in the case of materials they were expected to confine their experiments to proper engineering materials like wood and steel. Griffiths wanted a much simpler experimental material than wood or steel and one which would have an uncomplicated brittle fracture, for these reasons he chose glass as what is now called a 'model' material. In those days models were all very well in the wind tunnel for aerodynamic experiments but, damn it, who ever heard of a model *material*?

These things being so, Griffith and Lockspeiser took care not to bring the details of their experiments too much to the notice of the authorities. The experiments, however, involved drawing fibres and blowing bubbles of molten glass and one day, after the work had been going on for some months, Lockspeiser went home leaving the gas torch used for melting the glass still burning. After the inquiry into the resulting fire, Griffith and Lockspeiser were commanded to cease wasting their time. Griffith was transferred to other work and became a very famous engine designer. The feeling about glass died hard. Many years later, about 1943, I introduced a distinguished Air Marshal to one of the first of the airborne glass-fibre radomes, a biggish thing intended to be

---

Indiarubber has a breaking strain of about 700 per cent, but this depends upon quite a different mechanism which, as far as this calculation is concerned, may be regarded as cheating, see Appendix. The 'plastic' breaking strain in soft metals (say 60 per cent in mild steel) is not elastic, see Chapter 4.

bolted under a Lancaster bomber. 'What's it made of?' 'Glass sir.' 'GLASS! – GLASS! I won't have you putting glass on any of my bloody aeroplanes, blast you!' The turnover of the fibreglass industry passed the £100,000,000 mark about 1959 I believe.

To return to Griffith's experiments, Griffith was not the first man to draw strong glass fibres but he was probably the first man to do it in a systematic way and to provide a plausible explanation of the results.

Griffith had first to determine, at least approximately, the theoretical strength of the glass he was using. The Young's modulus was easily found by a simple mechanical experiment and two or three Ångström units is a fair guess for the interatomic spacing and cannot be far out.* It remained to measure the surface energy. It was here that one of the advantages of glass as an experimental material lay. Glass, like toffee, has no sharp melting point but changes gradually, as it is heated, from a brittle solid to a viscous liquid and during this process there is no important change of molecular structure. For this reason one might expect there to be no large change in surface energy between liquid and solid glass so that surface tension and therefore surface energy, measured quite easily on molten glass, ought to be approximately applicable to the same glass when hardened. When the end of a glass rod is heated in a flame the glass softens and tends to round off into a blob because surface tension remains active long after permanent mechanical resistance to deformation has disappeared. The force, which is easily measured, needed slowly to extend the rod under these conditions is therefore that which will just overcome the surface tension. From experiments of this type, done with very simple apparatus, Griffith could deduce that the strength of the glass he was using (at room temperature) ought to be nearly 2,000,000 p.s.i. or about 14,000 MN/m².

Griffith then took ordinary cold rods of the same glass about a millimetre thick and broke them in tension, finding that they had a tensile strength of about 25,000 p.s.i. or 170 MN/m² which is round about the average for laboratory glassware, window panes, beer bottles and most of the other common forms of glass but was

---

* As a matter of fact Griffith came at this calculation in a rather different way.

something between a fiftieth and a hundredth of what he reckoned it ought to be.

Griffith now heated his rods in the middle and drew them down to thinner and thinner fibres which after cooling he also broke in tension. As the fibres got thinner so they got stronger, slowly at first and then, when they got really thin, very rapidly. Fibres about one ten thousandth of an inch (2·5 mμ) thick showed strengths up to about 900,000 p.s.i. or 6,000 MN/m^2 when they were newly drawn, falling to about 500,000 p.s.i. or 3,500 MN/m^2 after a few hours. The curve of size against strength was rising so rapidly (Figure 1) that it was difficult to ascertain a maximum or upper limit to the strength. The increase of strength with thinness was not entirely smooth but showed a certain amount of scatter or

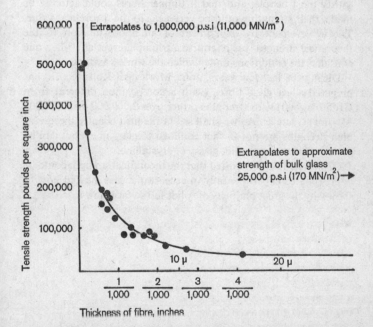

Figure 1. Griffith made and tested thinner and thinner glass fibres. As the fibres got thinner so they got stronger until the strength of the thinnest fibres approached the calculated theoretical strength.

variability. However, there was absolutely no doubt about the general trend.

Griffith could not prepare or test fibres thinner than about a ten thousandth of an inch (2·5 mμ) and, if he had, it would have been difficult at that time to measure the thickness with any sort of accuracy. However, by the simple mathematical device of plotting reciprocals it was possible to extend or extrapolate the size-strength curve fairly reliably so as to ascertain the strength of a fibre of negligible thickness. This turned out to be 1,600,000 p.s.i. or 11,000 MN/m^2. It will be remembered that Griffith had calculated a value a little under 2,000,000 p.s.i. or 14,000 MN/m^2 for the glass he was using. He therefore concluded that he had approached the theoretical strength quite closely enough to satisfy most people, and that if thinner fibres could actually be made, their strength would be very near to the theoretical value. The achievement by experiment of an approximation to the theoretical strength was of course a triumph, especially when one considers the conditions under which the work was done.

During the last few years, John Morley, of Rolls Royce, has prepared silica glass fibres (with a composition different from Griffith's glass) with strengths rather over 2,000,000 p.s.i. (14,000 MN/m^2) (Plate 4). As we shall see in the next chapter these very high strengths are not in fact confined to glass fibres but can be got from almost any solid, glassy or crystalline.

Griffith had demonstrated that the theoretical strength could be approximated experimentally in at least one case, he had now to show why the great majority of solids fell so far below it.

Chapter 4   Cracks and dislocations

or why things are weak

'The fault that leaves six thousand ton a log upon the sea.'
Rudyard Kipling, 'McAndrew's Hymn'.

Griffith wrote a classic Royal Society paper about his experiments which was published in 1920. In this paper he pointed out that the problem was not to explain why his thin fibres were strong, since a single chain of atoms must, inescapably, have either the theoretical strength or none at all, but rather to explain why the thicker fibres were weak.

It was becoming clear, at any rate to Griffith if to nobody else, that in a world where practical materials only reached a small and highly irregular fraction of the strength of their chemical bonds, the weakening mechanism, rather than the bond strength, was what really controlled mechanical strength. It is only quite lately and now that we are able regularly to get strengths which are a large fraction of the theoretical value, that it has become really important and worthwhile to make materials with very strong chemical bonds.

The weakness of glass fibres brings us to the question of Griffith cracks and it also brings us back to Professor Inglis, whom we left in Chapter 2 worrying about why ships broke in two at sea when simple calculation showed them to be amply strong enough. Inglis made calculations about the effect of hatchways and other openings in large structures like ships. Griffith had the wit to apply Inglis's mathematics on a far finer scale, to 'openings' of almost molecular size and too fine to see with an optical microscope.

Stress concentrations

Whatever the scale, the practical importance of stress concentrations is enormous. The idea which Inglis expounded is that *any* hole or sharp re-entrant in a material causes the stress in that material to be increased locally. The increase in local stress, which can be calculated, depends solely upon the *shape* of the hole and has nothing at all to do with its *size*. All engineers know about stress concentrations but a good many don't really in their hearts believe in them since it is clearly contrary to common sense that a tiny hole should weaken a material just as much as a great big one.* The root cause of the Comet aircraft disasters was a rivet hole perhaps an eighth of an inch in diameter. Small holes and notches are particularly good at starting fatigue failures but they also do very well for starting ordinary static fracture. When a glass cutter wants to cut glass, he does not bother to cut right through but makes a shallow scratch on the surface after which the glass breaks easily along the line of the scratch. (By the way, so-called 'cut glass' is ground to shape, not cut.) The weakening effect of the scratch has very little to do with the amount of material removed, a shallow scratch will do nearly as well as a deep one, it is the sharpness of the re-entrant that increases the stress.

It is not difficult to form a physical picture of what is actually happening at a re-entrant such as a crack, especially if we consider the matter upon a molecular scale. Referring to Figure 1 it is obvious that a single chain of atoms in tension must be uniformly stressed and should reach the theoretical strength (1a). The mere multiplication of such chains, side by side, to constitute a crystal, does not prevent each separate chain from still carrying its full theoretical stress (1b). Suppose now that we cut a number of adjacent bonds so as to constitute a crack, then of course we have interrupted the flow of stress in the broken chains and the load in these broken chains has got to go somewhere (1c). In fact it does the most natural thing, which is to go round the end of the opening. Thus the load in the whole of the cut chains may well have to pass through the single bond which closes the tip of

---

* As we shall see in Chapters 5 and 9, in a ductile metal the anelastic or plastic behaviour of the metal around a small hole may smooth out the local concentration of stress and greatly reduce the weakening effect of things like rivet holes; however, this is not always the case in fatigue, that is under repeated loading.

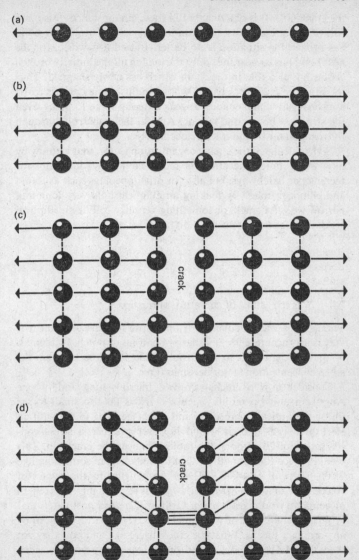

Figure 1. How a stress concentration arises at the tip of a crack.

the crack (1d). It is obvious that in these circumstances this bond will fail long before its companions. When this over-tried bond has broken the situation is no better. Indeed it is worse, for the next bond has to bear, not only the load in all the initially broken chains but also that in the chain which has newly snapped. Thus the situation goes from bad to worse. In this way a crack is really a mechanism which enables a weak external force to break even the strongest bonds one by one. And so the crack runs through the material until total fracture occurs.

Inglis calculated the stress concentration factor, that is to say by how many times the stress is increased locally, not only for rectangular hatchways, but also for other openings such as round and elliptical holes. By making an elliptical hole very long and narrow we get a crack, or something very like it. For an elliptical crack Inglis's stress concentration factor is:

$$\left(1+2\sqrt{\frac{L}{R}}\right)$$

where L is half the length of the crack
  R is the radius of curvature of the tip.

Though this was calculated for an ellipse it turns out that it is very nearly accurate for any sharp re-entrant or notch. Incidentally, for a circle, that is to say a round hole, it will be seen that the stress concentration factor becomes three.

Consider now a crack, say, two microns long and of one Ångström unit tip radius. Such a crack is far too small to see under the optical microscope and might be hard to see with an electron microscope. It would however produce a stress concentration of 201 among the molecules near the crack tip. This would reduce the strength of Griffith's glass from something like 2,000,000 p.s.i. or 14,000 MN/m² to a figure in the region of 10,000 p.s.i. or about 70 MN/m², which is very roughly the strength of common or domestic glass. Griffith therefore postulated that ordinary glass is full of very fine cracks, too small to be seen by any ordinary means. What the cracks were like and how they got there he did not say, but he did show that, if they existed, which was not unreasonable, they would account for the weakness of ordinary glass. He supposed that for some reason they became rarer in thin fibres and almost non-existent in the very thinnest fibres, perhaps because there was simply no room for them.

Griffith cracks

Griffith seems to have supposed that the cracks which he thought existed were scattered throughout the interior of the glass and were perhaps a consequence of a failure of the molecules inside the glass to join up completely when the glass hardened. Looking back, it is curious how long it took us to get rid of this idea of some kind of defect inside the body of the glass.

Griffith's calculations showed that the cracks, whatever they were, must be quite narrow, perhaps a hundredth of the wavelength of ordinary visible light. Since one can never hope to see things which are much smaller than the wave-length of the light which one is using, there was clearly no hope of ever seeing them directly by means of the ordinary optical microscope which reaches its limit with objects about half a micron thick. This is the reason for the invention of the electron microscope which uses electrons with a wave-length of something like a twenty-fifth of an Ångström instead of light with a wave-length of about four thousand Ångströms.

However, before the electron microscope was available, in fact in 1937, Andrade and Tsien decided to look for the cracks by the method of decoration, using of course an optical microscope. This method, which is often very effective, may be thought of like this. Imagine a fine wire which is too far away and too thin to see by any ordinary means. If we can persuade birds to come and perch on it in a row fairly close together we have obviously made it much easier to see the wire. (The Post Office sometimes put corks on the telegraph wires for the same reason.) If we can now persuade more birds to come and perch on top of the first lot we can in principle build up the wire to any thickness we choose. Now it happens that when some substances crystallize they find it easier to do so if some kind of irregularity is present. By choosing the right substance to crystallize on a surface one can often get the new crystals to form almost entirely upon the fine irregularities of the surface and thus show them up.

Andrade treated the surface of glass with sodium vapour and produced linear patterns which seemed to indicate the existence of surface cracks. In this work it is extraordinarily easy to produce ghost images and chimeras, like the canals on Mars, which may or may not be cracks, but even if Andrade's patterns showed genuine surface cracks, which seems likely, this would not prove that there were not also internal cracks.

In the years after the last War it was found that not only were the thinnest fibres strong, but, if they were carefully made, quite thick fibres could also show high strength. Strong fibres were weakened by touching while weak fibres were improved in strength if the surface were removed chemically.

All this was a strong indication that the important weakening defects lay on the surface and around 1957 Margaret Parratt, David Marsh and I spent a lot of time examining the surface of glass. By refining Andrade's sodium technique, Mrs Parratt was able to produce the most beautiful crack patterns on the surface of all kinds of glass and what was more, many, perhaps most of them, did seem to be genuine cracks. Furthermore the frequency of the cracks correlated quite well with the experimental strength of the various specimens of glass. The question was how the cracks got there. In many cases there was no doubt at all: the glass had been in contact with some other solid and the cracks were due to simple scratching or scraping. Plates 5 and 6, Mrs Parratt's photographs, show quite typical scrapes. Very little glass is wholly untouched from the time it is drawn or blown from the melt, and it takes only the lightest contact to create an elaborate crack pattern.

Very probably this simple explanation accounts for the weakness of the majority of common glass.* The high strength of thin fibres may be due in part to the fact that such fibres are very

---

*The Griffith energy criterion which governs the propagation of cracks is dealt with in Chapter 5. Though internal defects, where they exist, are not different in kind from surface ones, they are generally shorter and thus, as we shall see, may not have an energy incentive to propagate cracks.

easily bent and it is therefore easier to bend them than to scratch them. However, there do remain a number of cases where there are variations in the strength of glass whose surface is genuinely untouched. One reason for this was investigated by Marsh.

When most liquids freeze they crystallize and usually the crystal, being more orderly, is better packed and therefore more dense than the mother liquid. Water is an exception for complicated reasons. Glasses behave as they do because, while they are cooling, they are so viscous that the molecules do not have time to sort themselves out into crystals and so cool glass is a solidified liquid, not a crystalline solid. However the tendency to crystallize is there and given time some glasses will in fact crystallize. This is known as devitrification. Since devitrification involves shrinkages, the glass is often weakened and sometimes falls to pieces in the process. Devitrification is almost universal in ancient glasses which were usually badly made in the first place and have had plenty of time to crystallize; the result, however is, often very beautiful, though these old glasses have become very weak.

Marsh showed that in some glasses there is incipient devitrification even when they are new. He photographed tiny crystallites in the electron microscope and showed that the shrinkage which accompanied their formation was sufficient to initiate a crack which would spread into the main body of the glass (Plate 7).

It must be emphasized that there is nothing very special about thin glass fibres which is specifically due to their thinness as such. If the surface of thick glass can be got smooth and kept smooth it will be just as strong as a thin fibre. In practice, however, this is generally more difficult to do.

If a material like glass does not fail owing to the spread of a crack from some local defect then how does it fail? The answer is that it fails by flowing in shear, just like Plasticine or soft metals. Because the flow stress of glass is very high at room temperature and because glass is very susceptible to fracture by the spread of cracks, glass, and materials like it, nearly always fracture in the familiar brittle manner and we find it difficult to imagine anything different happening. In fact, if glass is prevented from cracking in tension, say by putting it into compression, then it is quite easy to get it to flow like a soft solid; for instance, glass will behave like putty under the blunt point of a diamond indenter but the shear stresses required to cause flow are well above the normally observed fracture stresses – in common glasses at room temperature usually above 500,000 p.s.i. or 3,500 MN/m².

Fairly recently Marsh has shown that glass which is quite free from cracks does in fact fail in this manner by flowing, and that when this happens around room temperature, the stresses are usually upwards of 500,000 p.s.i. An interesting point is that the tendency to fracture by the spreading of cracks is relatively little affected by temperature whereas the viscosity or shearing stress is very dependent upon temperature. For this reason, when we heat glass to a temperature well below its melting point the shearing stress is reduced more than the brittle fracture stress and thus we can bend and shape and blow hot (but not necessarily very hot) glass quite easily. Contrariwise, defect-free glass can be strengthened by cooling which raises its viscosity or resistance to flowing. In this way the strength of smooth glass tested at −180°C. is about twice that of the same glass tested at room temperature.

Very generally, there are always two fracture mechanisms competing to break a material – plastic flow and brittle cracking. The material will succumb to whichever mechanism is the weaker; if it yields before it cracks the material is ductile, if it cracks before it yields it is brittle. The potentiality of both forms of failure is always present in all materials.

The strength of brittle crystals and the whisker story

All this accounts fairly satisfactorily for the strength and weakness of glasses, with which are included natural glassy minerals such as flint* and obsidian, but then the vast majority of hard solids, both natural and artificial, are crystalline. There is some kind of popular superstition that crystalline materials are weak. The garage foreman, appearing with your broken crankshaft or back axle, will tell you that it has 'crystallized'. What state it was in

* Strictly speaking flint is micro-crystalline but, mechanically, it behaves very like a glass.

before it crystallized he does not explain; certainly it was not glassy. Needless to say all metals, the great majority of minerals, most ceramics and common solids like salt and sugar are crystalline. From common sense one would not expect the mere possession of a regular, orderly arrangement of atoms or molecules to be a cause of weakness, and of course it isn't.

When we are dealing with hard, brittle crystals, however, the practical strength is generally even lower than that of bulk glass and in their crude state most of the non-metallic crystals deserve the contempt with which they are generally regarded by engineers.

At this stage it is necessary to talk about whiskers. People often mention 'metal whiskers' as if they were the only kind but, as a matter of fact, metal whiskers are less common and less interesting than whiskers of non-metals and it is about these latter that we shall mostly talk. Whiskers have nothing to do with human hair and are in fact long thin needle crystals which can be grown from most substances by accident or by care. They can be grown in a large number of different ways but are typically one or two microns thick though they may be millimetres or even centimetres long.

Whiskers sometimes grow by accident from the surface of metals and if the metal happens to be part of an electrical device then there is likely to be a short circuit which will be annoying, expensive or dangerous according to the circumstances (Plate 9). Metal whiskers of this kind had been known of, in a general sort of way, for a long time but had been regarded as a nuisance or a curiosity. They were not thought particularly interesting, until, in 1952, Herring and Galt chanced to bend some tin whiskers and noticed that they could be bent to a strain of about 2 per cent and still recover elastically. This corresponded to a higher stress than had ever been observed before in tin and perhaps in any other metal. It looked like another case of thin fibres showing anomalously high strength and it naturally attracted a good deal of attention.

Herring and Galt worked with tin. Tin is a metal and somehow everybody expects metals to be strong. What interested me at the time was whether the 'naturally' weak non-metallic crystals could be made strong too. Thinking about this one morning in 1954, I went along to the keeper of the laboratory chemical stores and asked him for something which was water-soluble and formed needle crystals. He gave me a bottle of hydroquinone, a common substance used in photographic developers. The bottle was full of dry crystals about as thick as an ordinary pin and about a centimetre long. Manipulating these crystals by hand with the dissecting tools which biologists use it was quite obvious that their strength was negligible. I then dissolved some of the hydroquinone crystals in water, put a drop of the solution on an ordinary glass microscope slide and allowed the water to evaporate naturally in air so that new, but much smaller, needle crystals were produced as the solution dried under the microscope.

The new crystals tended to be long, smooth, whip-like filaments, initially so thin that they were hardly visible in the optical microscope. By poking about with a dissecting needle it was obvious that these little threads were very strong, exactly how strong it was difficult to say (Plate 12). This was exciting and I was very soon trying crystals of all sorts of substances taken from the shelves of my own and my colleagues' laboratories. With a little skill and low cunning it was possible to get almost any common soluble solid, such as Epsom salts, or even sodium chloride – ordinary table salt – to crystallize in the form of these very thin filaments, whiskers, and in every case these whiskers were obviously strong. It might be supposed that their strength had something to do with the crystals being wet. A man called Joffé observed, about 1928, that some things got stronger when they were wet. (Actually, other things get weaker.) However, drying the whiskers out did not seem to have an important effect upon their strength, as far as I could tell.

One of the difficulties in the early stages of this work was to find any reasonably reliable method of measuring whisker strengths. We used to bend the whiskers under the microscope with a dissecting needle and, having measured the thickness and radius of curvature very approximately we could estimate the breaking strain by simple beam theory. As one might imagine this method was maddeningly inconvenient and very inaccurate.

The whiskers generally began life as exceedingly fine filaments which could be seen to thicken as they got older. I therefore introduced the refinement of bending the infant whisker, by troubling the waters around the moment of its birth, and then simply allowing it to thicken until it broke, this was a little less clumsy but still highly unsatisfactory.

Just at this time (1956) David Marsh came to work with me and one of the first things he said was 'Why don't you build a proper tensile testing machine?' I am afraid that I told him to go away and not be silly. The whiskers were too small to see with the naked eye and nobody could possibly make a testing machine on that scale. Marsh went away and was not silly for he came back with a micro-testing machine which worked, which he had designed and built himself. The Mark III version of the Marsh machine went into commercial production and now no gentleman's laboratory is complete without one. This remarkable machine will, if pressed, test fibres down to one tenth of a square micron cross-section (that is, virtually invisible in the optical microscope) and about a quarter of a millimetre long. It will detect extensions down to less than five Ångström units which is about the resolution of a good electron microscope.*

With this tool in our hands, we were able to get some real results. The first thing we discovered was that we could get high strength from almost anything from Epsom salts to sapphire. Provided it was in the form of a thin whisker, it did not matter what the chemical nature of the stuff was or by what method the whiskers were grown. We must have worked on well over a hundred different substances and there was absolutely no doubt about it.

When we plotted strength against thickness for any given whisker we got a curve which was uncannily like Griffith's size-strength curve for glass fibres (Chapter 3). What was more, when we plotted, not strength but breaking strain, against thickness, we found that it did not matter what the whiskers were made of, for they all plotted on the same curve. Figure 2, for instance, shows the breaking strains of whiskers of two very different substances, silicon and zinc oxide. It is impossible to tell them apart.

*The construction and working of this machine is described in the *Journal of Scientific Instruments* for 1961 (D. M. Marsh, 38, 229–34.)

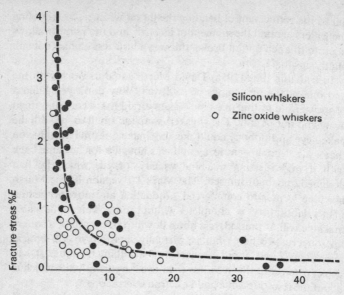

Figure 2. Strength-diameter relation for whiskers. Like glass fibres, they get stronger as they get thinner, but the cause of the increase of strength is different.

The temptation to assume that the strength and weakness of whiskers, and therefore presumably of other crystals, was due, like glass, to surface cracks was of course strong. However, we were unable to find any cracks and there were good reasons, based on the mode of growth, to suppose that they should not exist. When a whisker grows from solution or from vapour there generally first appears a very fine filament or leader which in the electron microscope can be seen to be almost perfectly smooth. This leader then thickens by the spreading down it of new, sleeve-like layers of material. At first these layers may consist of single layers of atoms or molecules, but, naturally, the various successive layers of new material will be fed with molecules from the surroundings at slightly differing rates. Thus the layers which are

fed faster will advance faster and may catch up on slower underlying layers which they have no means of passing. The advancing
edge or step is now twice the normal height and so would require
twice the amount of new material if it were to advance at the same
rate as the other layers. In fact, however, it is only likely to be
supplied by diffusion at the same rate as the single layers. Thus
the double layer moves more slowly than the average and so yet
more growth layers pile up behind it and cannot pass. Thus there
build up a series of steps with sharp, cliff-like fronts (Plate 10). On
the average these steps will be higher the older and therefore the
thicker the crystal. When the crystal stops growing, because it is
removed from the solution, or for any other reason, these steps
will remain on the surface and can often clearly be seen in the
microscope.

It is quite easy to see, intuitively, that a crack is a nasty dangerous thing to have about the house but it is by no means so obvious
that a step can cause a bad stress concentration. The problem of
the step was so little regarded that there was no standard solution in the literature and so I asked David Marsh to get one.
Working with a series of transparent resin models in polarized
light, Marsh was able to show that a step was just as bad a
stress concentrator as the equivalent crack, in fact it might be regarded as half a crack. This experimental solution has since been
confirmed by a purely mathematical analysis carried out by
H. L. Cox.

Though this work was done to explain the strength of very tiny
crystals it is worthy of the attention of engineers who are perhaps
somewhat frightened of cracks but take a rather light-hearted
view of steps in machinery and structures.

It will be realized that with the step, as with the crack, what
governs the stress concentration or weakening effect is not the
absolute size of the re-entrant but the ratio of depth to the root
radius.

Marsh examined a number of whiskers in the electron microscope and found that, for the substances he was using, the root
radii of the growth-steps was roughly constant at about 40
Ångströms. He then compared the heights of the worst steps with
the measured strength of the whiskers. The correlation was

extraordinarily good and left no room for doubt that this is the true explanation of the size-strength effect in whiskers. Since a large whisker is in no way different from any other crystal except in size it must also be a general explanation of the strength and weakness of all brittle crystals.

That the strength behaviour of whiskers is no different from that of much bigger crystals was confirmed by Dash who took a large (2 cm.) crystal of silicon, which is normally quite a weak material, and polished it very carefully. He then enclosed it in a transparent box which he provided with a mechanism for straining the specimen in bending. He used to take this affair round various conferences and demonstrate to all and sundry that the crystal could be repeatedly bent to a strain of 2 per cent which is equivalent to a stress of about 600,000 p.s.i. or 4,000 MN/m², a very respectable figure.

When we come to everyday crystalline materials, however, there is one more link in the argument. It is possible, by taking care, to grow quite large single crystals, as Dash did, but normally each individual crystal in a common material is quite small. Whiskers are small single crystals of rather a special kind but ordinary sizable solids are what is called polycrystalline, that is to say they are made up of a large number of small crystals fitting together in three dimensions like crazy paving or like counties on a map. Although the shape of the individual crystals is irregular the fit at the boundaries is usually very good and in a pure material there is good contact on a molecular scale. In general the surface energy of these boundaries is actually higher than that of fracture planes within the crystals and so, in a reasonably pure material, the 'grain-boundaries' are not usually a source of weakness.

The case is somewhat different however with an impure material. As is well known, when a liquid freezes by crystallizing the crystals tend to expel impurities. For instance ice which is formed on salt water is substantially fresh, to the great convenience of Polar explorers. This effect causes impurities in solids to accumulate at the grain boundaries (and also vacancies, that is, holes) and this may cause the grain boundary to become a line of weakness. This is one of the reasons why the addition of quite small amounts of the wrong impurity can ruin an alloy. A useful application of this weakening effect occurs when we add anti-freeze to the water in a car radiator. It is true that the glycol does depress the freezing point of the resulting mixture considerably and so postpones freezing but, when this eventually does happen, the result is a mushy ice without mechanical strength which is unlikely to do much harm to the engine.

For most fairly pure crystalline solids, however, the grain-boundaries are quite strong and, for a hard brittle material, a polycrystalline solid may be regarded as behaving in a manner comparable to whiskers and other single crystals and this, as we have seen, is very like the way glass behaves. In both cases the problem of strength and weakness is almost entirely a matter of surface smoothness. In the case of glass the important defect is usually the surface crack, in the case of brittle crystals it is usually the surface step. The presence of internal defects in a brittle crystal is usually of minor importance.

As we shall see, the problem with a ductile material, such as a soft metal, is quite different.

Dislocations and ductility

So far we have dealt entirely with what are technically known as 'brittle' substances. This does not mean, of course, that they will fall to pieces at a touch and as we have said, some brittle sub-stances are very strong. There is no absolutely sharp division between brittle and ductile substances but generally speaking brittle solids have fairly well defined characteristics. Apart from the small elastic strains which recover when the load is taken off, brittle solids do not distort before failure and fracture is usually by a crack or cracks which run cleanly through the material. Thus the bits will fit together after fracture so that one can often glue a broken vase together quite plausibly. In a ductile material such as mild steel, a good deal of irreversible distortion takes place before actual fracture occurs so that broken parts do not fit even approximately. This is one of the reasons for the high cost of car repairs.

Brittle substances in common use include glass, pottery, bricks, cement and some plastics and these are fairly satisfactory for the purposes for which they are generally used. For the more exacting uses, such as machinery, we generally tend to choose ductile metals. In a brittle solid, fracture occurs by the total separation of two adjacent layers of atoms or molecules under a tensile stress, the rest of the material being undisturbed. A metal behaves rather more like Plasticine. Before actual fracture occurs, in the sense that the specimen separates into two pieces, there is extensive flow, something like a viscous liquid, in the body of the material. This is caused by adjacent layers of atoms, not coming apart, but sliding over each other after the fashion of a pack of cards.

After the adjacent layers of atoms have slipped to a greater or less extent, so that the material is deformed in shear, no serious weakening has necessarily taken place, so that broken bonds have reformed with new partners. In some cases metals are actually strengthened by this process, which is known as cold working. If the process is carried too far the material will however be weakened and eventually broken. The amount of shearing or elongation which a ductile material will withstand varies enormously between different metals and alloys. It nearly always increases when the metal is heated; hence the village blacksmith and his forge. The ability of ductile metals to be permanently distorted and therefore shaped either cold or hot is of course an outstanding advantage of metals. Besides this, it goes a long way to account for their toughness, as we shall see in Chapter 9. However it is also the principal cause of their weaknesses. This is because, as we have said, if a material does not fail in a 'brittle' manner because of a crack at right angles to a tension stress, then it may fail by 'sliding off' (Figure 3) at 45° to it, and if this is the weaker mechanism this is what will happen.

Kelly has shown recently that correct calculations about the shearing strength of solids are rather complicated and that there is a good deal of variation between the theoretical shear strengths of various substances. However we can achieve an approximation to the theoretical shear strength by means of a very simple model and the result is not grossly in error. Consider a model, on paper or in the solid, consisting of sheets of spheres or balls to represent the atoms. There will be certain positions in which such

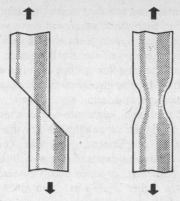

Figure 3. Ductile metals tend to fail in tension by shearing on planes at approximately 45° to the direction of the tensile stress. This generally leads to a local contraction or 'necking' of the material.

sheets of spheres will lie on each other as close together as possible. To disturb the layers from this position involves moving them further apart, which is resisted by tension in the bonds. The rows of atoms, as it were, dislike being dug out of their comfortable trenches of lowest energy.

Figure 4 represents a two-dimensional model such as is afforded by two parallel rows of pennies lying on a table. It is clear that all

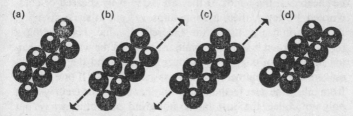

Figure 4. Shearing by sliding of whole planes of atoms without benefit of dislocation mechanisms.
(a) Initial position at rest.
(b) Resistance to slip a maximum, about 15° angular shearing displacement.
(c) All resistance to shearing gone at 30°.
(d) Final position of rest 60° shear.

resistance to shearing has vanished by the time that the atoms are balanced on top of each other, which will occur when the whole material has been distorted in shear through an angle of 30°. Beyond this point the rows of atoms will subside into repose in the next valley and shearing through one atomic spacing will have occurred. The resistance to this shearing will start at nothing, build up to a maximum, and decline again to zero when the atoms get to the top of the hump. Resistance will be a maximum about half-way up the hill, in this case about 15°. The three-dimensional case is slightly more complicated but the peak occurs around 7°. Crystals made from arrangements of atoms of differing sizes may also tend to reduce the angle at which maximum resistance occurs.

Arithmetic of a rather crude sort again gives figures in the general region of 10 per cent of the Young's modulus for the theoretical shear strength. (Kelly's more sophisticated approach gives figures between about 5 per cent and 10 per cent E.) In a way it does not matter very much if our figure is not very accurate because it is seldom or never reached when testing real materials in bulk.* The theoretical figure for iron is something like 1,500,000 p.s.i. or 10,000 MN/m² but in practice a crystal of really pure iron shears at about 3,000 p.s.i. or 20 MN/m², commercial steel at about 25,000 p.s.i. and the very strongest steel at 250,000 or so.

Really soft metals like pure gold, silver and base lead can easily be sheared in the hands. If they are extensively sheared or cold-worked, however, there is some improvement in shear strength, though not nearly up to the theoretical value. Hammering a metal to harden it is not uncommon: this was the way of hardening the edges of copper and bronze weapons and the old clock-makers always hammered their brass gear-wheels. (If one refrains from oiling the gear teeth of a grandfather clock, the teeth will not only not collect the dust and so not grind each other away, but also become harder and more polished as times goes on and so last virtually for ever.)

---

*It will be noticed that, very approximately, the theoretical shearing stresses are generally lower than the theoretical tensile stresses. Thus, near-perfect specimens tend to fail in shear if they can reach very high stresses, as we have seen in the case with glass.

Until about 1934 the Establishment explanation of these phenomena was remarkably unconvincing and seems to have reflected mainly a desire not to be asked embarrassing questions. 'Slip is due to little bits of crystal getting loose and acting as roller bearings between the layers. When too many break away they jam each other and that is the cause of work hardening.' As the Duke of Wellington said, 'If you believe that you will believe anything.'

In 1934 G. I. Taylor, a Cambridge don who invented the ploughshare anchor, a major alleviation of human misery, also invented the dislocation. At least he put up the dislocation as a hypothesis in a scientific paper. The essential idea is extraordinarily simple, so simple that it must be true, and it turns out that it is.

It is most unlikely, said Taylor, that metal crystals are really as perfect as we suppose them to be when we do sums about their strength. Let us suppose that every now and then, perhaps every million atoms or so, slight irregularities occur. What we need is not so much a point irregularity such as a foreign atom, because that could only facilitate movement at one point, but rather a line defect which will allow the army of molecules, as it were, to sweep forward on a broad front.

Crystals, of course, consist of sheets or planes of atoms, which to an electron-sized observer would seem to lie, piled upon one another, in awful and endlessly regular array, virtually for ever, like the pages of some enormous celestial book. What Taylor suggested was that every now and then, but very rarely, a sheet of atoms is not complete. It is as if somebody had slipped an extra sheet of paper between the pages of the book which now, at one point consists of perhaps a million pages, at another of a million and one. The interesting bit occurs of course along the line where the extra layer of atoms comes to an end.

Referring to Figure 5, it will be seen that there must be two places, one on either side of the tip of the extra sheet, where the atoms are distorted to an angle which approximates to the theoretical shearing strength of the crystal. In other words, at these two points the crystal is pretty well broken away.

What is even more important, the dislocation turns out to be

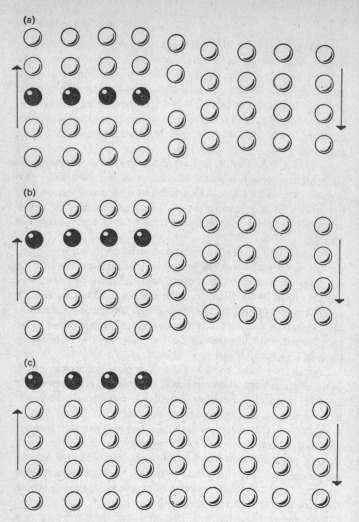

Figure 5. Shearing by means of an edge dislocation. For the reality of this phenomenon see Plate 14.

(a) Edge dislocation (diagrammatic).
(b) Dislocation sheared one lattice spacing.
(c) Dislocation sheared out of crystal altogether.
Note. The shaded atoms are not of course the *same* individual

movable. If we put a gentle shearing force upon the crystal as a whole we can easily apply that little extra strain needed to break the whole line of sorely stretched bonds but we find (Figure 5b) that we have merely reformed the whole arrangement one atomic spacing further on. By going on loading the crystal in shear we can repeat the process indefinitely and eventually squeeze the dislocation out of the far side of the crystal (Figure 5c). Furthermore the force needed to do so may be very small.

Engineers and some metallurgists resisted the idea with the whole force of their emotions and even today some of them are still making growling noises in caves in the backwoods. Academic physicists, on the whole, however, fell upon dislocations with glee. For many years nobody saw a dislocation in the flesh, or perhaps ever expected to, but their hypothetical movements (dislocations of like sign repel each other etc.) and breeding habits (when the union of two dislocations is blessed about five hundred new dislocations are suddenly released upon the crystal) could be theoretically predicted and provided a superb intellectual exercise like three dimensional chess.

As a matter of fact nearly all these academic predictions turn out to be true. Taylor supposed originally that slip in ductile crystals was due entirely to those dislocations which were present initially in the crystal due to the accidents of imperfect growth. It turns out that there are generally not enough dislocations originally present in most crystals to account for the very extensive slip which can take place in a ductile material. Large families of new dislocations can however be nucleated either by dislocation interaction (known as a Frank-Read source), or, more frequently, by severe stress concentrations, such as occur at crack tips. These mechanisms enable a stressed metal to be rapidly filled with dislocations (something like 10^{12} per square centimetre) and thus to flow under a steady load or the blow of a hammer quite easily.

---

atoms in each of these diagrams. The shading merely indicates the position of the extra half sheet of atoms. During dislocation movement no individual atom has to move more than a fraction of an Ångström from its original position.

It will be recalled that the dislocation is essentially a line defect which can move about in the crystal fairly freely. When there are many dislocations they do not have to move far before two or more dislocations meet. In rather special circumstances this can result in the creation of new dislocations but the much more usual effect is for them to repel each other. As more and more dislocations are born and move about they impede each other and get tangled up, like so much string. The result is that after a period of free movement the material begins to harden and if one goes on deforming it it will become brittle.

The most familiar example of this is when one wishes to break off a piece of metal such as a wire or the opened lid of a tin can. This can usually be done by bending it backwards and forwards a few times. The metal yields easily at first, hardens somewhat and then breaks off in a brittle fashion. Metal hardened by deformation can be returned to its initial soft condition by 'annealing', that is to say by heating it until total or partial recrystallization occurs, in which case most of the excess dislocations vanish. Thus copper tubes must be annealed after bending to shape or they will be brittle.

Altogether, the dislocation mechanism has been found to explain the mechanical properties of metals very well indeed. Although dislocations do exist in non-metallic crystals they are not usually very mobile and they seldom breed, thus dislocation movement does not play any important part in the way non-metals behave. It is the mobility of dislocations which accounts for the mechanical differences between metals and non-metals. Needless to say, dislocations cannot exist in glasses because glasses are not crystalline.

Part Two

The non-metallic tradition

or how to be tough

> '*And yet flint is considered to involve comparatively easy work, as there is a kind of earth consisting of a sort of potter's clay mixed with gravel, called "gangardia" which it is almost impossible to overcome. They attack it with wedges and iron hammers; and it is thought to be the hardest thing that exists, except the greed of gold which is the most stubborn of all things.*'

> Pliny, *Natural History.**

Pliny the Elder (A.D. 23–79) in his highly unreliable *Natural History* gives directions for distinguishing a genuine diamond. It should be put, he says, on a blacksmith's anvil and smitten with a heavy hammer as hard as possible; if it breaks it is not a true diamond. It is likely that a good many valuable stones were destroyed in this way because Pliny was muddling up hardness and toughness. Diamond is the hardest of all substances and the hardness of diamond is useful when one wants to cut or scratch or grind materials and this is its main industrial use. But diamond, like other hard precious stones, is quite brittle so that, even if one could get it cheaply in large pieces it would not be a very useful structural substance.

The worst sin in an engineering material is not lack of strength or lack of stiffness, desirable as these properties are, but lack of toughness, that is to say, lack of resistance to the propagation of cracks. One can allow for lack of strength or stiffness in design but it is much more difficult to allow for cracks which catch the engineer unawares and are dangerous.

Most metals and timbers, and also Nylon, Polythene, fibre-glass, bones, teeth, cloth, rope and jade are tough. Most minerals, glass, pottery, rosin, bakelite, cement, and biscuits are brittle, and so is ordinary table jelly as one can readily prove by propagating a crack in it with a spoon and fork. It is not at all easy to see what it

*Translated by Racham and Jones.

is that makes one thing tough and another brittle because the substances in each of these lists seem to have little enough in common. The distinction is a very real one however. Pottery and commercial tinplate may have roughly the same tensile strength but if a cup is dropped on the floor it will shatter, perhaps almost explosively. If we drop a tin can probably nothing will happen; at the worst we may make a small dent. The actual tensile strength of ordinary glass and ceramics can be quite high; the reason why we do not make motor cars, for instance, from them is not that they are weak but that they are far too brittle. Anybody can tell this from common sense. But why? What is really happening?

First of all brittleness is not primarily a matter of the rate of loading. Psychologically there is a great difference between a statically applied load, that is one which is put on slowly, and a dynamic load, that is one suddenly applied, usually by means of a blow. The distinction does arise and it cannot entirely be neglected but it is much less important than appears at first sight. Generally we use a hammer simply because it is a convenient and cheap way of getting a high local force, which we do by decelerating the heavy head. Usually we should get much the same results by applying the same load slowly. To a certain extent this applies to dropping things on the floor and to car and aeroplane crashes, though as we shall see there are some important reservations. However, in most brittle materials, whether the force which causes fracture is applied quickly or slowly, once failure has begun, the consequent cracks will propagate very quickly indeed, usually at several thousand miles an hour. Thus to the eye fracture appears to be instantaneous.

In a way there is no essential difference between a stressed material and an explosive. When an elastic material is strained, strain energy is stored in the stretched chemical bonds and when the material is fractured this energy is released. At the theoretical breaking strain of the material all the bonds are stretched to their maximum and the strain energy is roughly equal to the chemical energy, as we ought to expect. In practice materials generally only reach a small fraction of their theoretical strength before they break and so the release of strain energy is usually far less than the energy which would be provided by an equivalent weight of explosive. All the same, it may make a very respectable bang. When strong fibres and whiskers are broken, for instance in the Marsh machine, one can realize a high proportion of their theoretical strength. In these cases one is not left after fracture with two or more broken pieces: there is an explosion and the fibre vanishes in fine dust. Only the fact that these strong fibres are generally quite small prevents the operation of breaking them from being a dangerous one.

Impact strength

Before going on to the general question of crack propagation and control, it is worth considering some of the special effects of a dynamically applied load such as a blow. The highest speed at which a stress can be transmitted through any substance is usually the speed of sound in that substance. Indeed, sound is perhaps best thought of as a wave, or series of waves, of stress passing through a substance at its natural speed.

Now the speed of sound in any substance is $\sqrt{E/\rho}$ where ρ is the density or specific gravity of the substance and E is the Young's modulus. Given the common values of E and ρ in structural solids we find that the speed of sound in these substances is very high indeed: for steel, aluminium and glass it is about 11,000 miles an hour or 4,800 metres per second, which is much faster than the speed of sound in air. Such speeds are far faster than any hammer blow and considerably faster than the flight of bullets.

The result is that the hammer or the bullet is pressing against its target for a period, perhaps about a hundredth of a second, which is very long compared with the time which is required to conduct the energy away from the point of impact in the form of waves of sound or stress. As photographers know, a great deal can happen in periods of time as long as a hundredth of a second. What is apt to happen when we strike a solid is that a whole series of stress waves radiate from the point of impact and move off into the body of the material. They reach the further boundaries of the solid in a time which is probably between a ten-thousandth and a hundred-thousandth of a second and are reflected back, as a kind of echo, very little attenuated or diminished in intensity. What happens next depends upon a great many things such as the shape of the solid, exactly where the blow was struck and so on. What *may* happen is that the returning reflected stress waves repeatedly meet the outgoing ones at some critical or unlucky point and thus the stress may pile up at this point until fracture occurs. The stories about singers fracturing panes of glass may well be true.

There are some elegant instances of the sort of things which can happen. The British Ceramic Research Association, for instance, have a routine impact test for ceramic tiles in which a loosely supported square tile is struck a measured blow in the centre of one flat face. In many cases the tile does not break in the middle where it is struck. What happens is that the four corners drop off because the stress waves are reflected and crowded into the corners.

When a shell is fired against armour plate, if the shell does not penetrate, then what is known as a 'scab', a jagged piece of armour, sometimes becomes detached from the inside or back surface. If this happens the scab may bounce about with great speed and energy within the turret and may do as much damage as if the shell had actually penetrated.

In a similar way, when a projectile is fired into a tank of liquid, such as the fuel tank of an aeroplane, it is the exit hole which is much the largest and the most difficult to seal since the shock waves which are readily transmitted through the fluid may burst the back of the tank. Unfortunately the human head is structurally rather like a tank of liquid and the consequences when it is struck by a bullet are well-known. What is less well-known is that rather similar effects may happen with a blow on the forehead which does not penetrate. The important factor in the design of crash-helmets is therefore the cushioning of the shock wave so as to prevent damage at the back of the skull. This is the reason for the internal head-band in helmets which looks as if it was put in in order to provide ventilation.

In engineering it is usual to test materials for toughness by means of a routine impact test of what is called the 'Izod' type. In this test the material is in the form of a bar a quarter or a half an inch square and is often provided with a standard notch to initiate failure. This bar is clamped at one end and the other end is then broken off by means of a heavy hammer in the form of a pendulum. By measuring the difference in the height to which the pendulum swings before and after breaking the specimen the energy of fracture can be estimated. Academically, this test is not very accurate but it does have some value as a rough comparative test between different materials.* It is very popular with engineers.

The Griffith criterion and critical crack length

To return to crack propagation in brittle solids, it does not really matter for our present purposes whether fracture is initiated by a dynamic blow or by a static load. On the whole, if the fracture stress is produced, by whatever means, at a given point, fracture will probably occur there. There are some exceptions to this: a few substances, such as pitch and toffee, are sensitive to the rate of loading. Every child knows that the way to break the most intractable toffee is to hit it with a poker. This will work when slower methods are quite impotent (Chapter 9). However, most normal materials are not much affected by considerations of dynamic and static loading.

Of course it would be ideal to have a material in which it was impossible to initiate cracks at all. Unfortunately in practice this does not seem to be a possibility. As we have seen in the last chapter the surface of even the smoothest glass is infested with tiny invisible cracks and even if it were not, it soon would be when it had brushed against some other solid. What therefore counts most of all is the ease with which the cracks can be made to extend by applying a stress to the material. The basic theory of crack propagation is again due to A. A. Griffith.

Griffith said that two conditions must be fulfilled if a crack was to propagate. First it must be energetically desirable and secondly there must be a molecular mechanism by which the energy transformation can take place. The first condition requires that at every stage in the propagation of the crack the energy stored in the material is being reduced, just as when a car runs

* As we shall see on page 108, it is an approximate measure of the 'work of fracture'.

downhill its potential energy is being reduced. On the other hand, however energetically desirable it may be, the car will not run downhill unless it is provided with wheels and the brake is off. The wheels are the mechanism by which the car runs downhill and implements the transformation of energy.

As we have said, a strained material contains strain energy which would like to be released just as a raised weight contains potential energy and would like to fall. If the material is completely fractured naturally the whole of this energy is in the end released. Consider, however, what happens during the intermediate stages of fracture. When a crack appears in a strained material it will open up a little so that the two faces of the crack are separated. This implies that the material immediately behind the crack is relaxed and the strain energy in that part of the material is released. If we now think about a crack proceeding inwards from the surface of a stressed material (Figure 1) we should expect the area of material in which the strain is relaxed

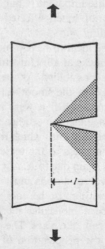

Figure 1. Griffith criterion for crack propagation. As the crack extends the material in the shaded areas is relaxed and releases its strain energy. This released energy then becomes available to propagate the crack still further.

to correspond roughly to the two shaded triangles. Now the area of these triangles is roughly l^2, where l is the length of the crack. The relief of strain energy would thus be expected to be proportional to the square of the crack length, or rather depth, and in fact this rough guess is confirmed by calculation. Thus a crack two microns deep releases four times as much strain energy as one one micron deep and so on.

On the other side of the energy account book is the surface energy, $2G.l$, which is needed to form the new surfaces and clearly this increases only as the first power of the depth of the crack. Thus a crack two microns deep has twice the surface energy of a crack one micron deep but as we have seen it releases four times as much strain energy. The consequences of this are fairly clear. When the crack is very shallow it is consuming more energy as surface energy than it is releasing as relaxed strain energy and therefore the conditions are energetically unfavourable for it to propagate. As the crack gets longer however these conditions are reversed and beyond the 'critical Griffith length' l_g the crack is producing more energy than it is consuming, so it may start to run away in an explosive manner. There is a characteristic critical Griffith crack length for each stress in the material. At the theoretical maximum stress the critical crack length l_g is extremely short and at zero stress it is infinitely long, which is what we should expect.

Algebraically, this can be shown to be equivalent to,

$$l_g = \frac{2GE}{\pi s^2}$$

where l_g is in metres

G is in J/m^2

and E and s are in Newtons/m^2 (*not* MN/m^2)

If the energy required to propagate the crack by producing the two new fracture surfaces were really confined to G, the free surface energy (which is seldom much above 1 J/m^2), then simple arithmetic will show that, at any realistic stress level, the critical crack length would be very short indeed, perhaps around a micron.

Fortunately this is seldom actually the case with practical

materials because, in order to produce a new fracture surface, we have generally not only to break all the chemical bonds at the fracture surface (which requires only the free surface energy), we also disturb the molecular structure of the material to a depth which is sometimes very considerable; in doing so we break a great many other bonds as well. In other words, fracture is a brutal process. The total energy which has to be used up in producing a real fracture surface is therefore greater than G and is known as the 'work of fracture', W, such that,

$$l_g = \frac{2WE}{\pi s^2}$$

– W being in J/m^2 again.

Although W is always considerably bigger than G, its actual magnitude varies very greatly between different materials. With glass, for instance, the molecular structure is only disturbed to a comparatively shallow depth below the fracture surface and W is generally around 6 J/m^2 – in other words about six times G – and so, although l_g, the critical crack length, is six times as high as it otherwise would be, it is still very short and glass is a brittle material. Most ceramics are not much better than glass in this respect but the ductile metals, such as wrought iron, mild steel, copper and aluminium, have works of fracture which are enormously higher than their free surface energies and range between 10^4 and 10^6 J/m^2. That is to say, W is from ten thousand to a million times higher than G and so the critical crack length is longer in direct proportion. Thus mild steel structures, for instance, can generally put up with cracks at least a metre long without breaking. This is what makes ductile metals so safe and tough and so popular.

In fact the possession of a high work of fracture – at least a thousand times higher than the free surface energy – is an essential characteristic of all safe and practical structural materials which are used in tension. However the actual molecular mechanism by which so much energy can be absorbed during fracture varies a great deal between the different kinds of solids. With metals it is essentially due to the operation of the dislocation mechanism which was described in the last chapter; we shall discuss how this works in Chapter 9. Timber, again, has a high work of fracture – about 10^4 J/m^2 – but this is produced by a totally different mechanism which we shall talk about in Chapter 6. The various plastics and composites are different again and we shall come to these presently.

As we saw in Chapter 4 the stress concentration at the tip of a crack is about:

$$K = 2\sqrt{\frac{l}{R}}$$

where l = crack length, R = tip radius.

Note L is the half length of an internal (or elliptical) crack, l is the length of a crack proceeding inwards from a surface. For all practical purposes L may be taken as equivalent to l.

Now in many materials, R, the tip radius of the crack, remains constant whatever the crack length, so that as the crack gets longer, the stress concentration gets worse. In practice R may have a value comparable to atomic dimensions, say about 1 Ångström unit. A crack length of about 1 micron (i.e. 10,000 Ångströms) will therefore produce the theoretical stress at the crack tip when the average stress in the material is quite modest and for longer cracks the stress at the tip will be far higher. However, paradoxically, and as Griffith pointed out, *as long as the crack is shorter than the critical length, nothing will happen – however high the stress at the tip may be.* Thus the chief safeguard against brittle failure lies in a high work of fracture. This is basically why glass ($W = 6$ J/m^2) is brittle and steel (W = about 10^5 J/m^2) is tough, although both these materials have roughly the same tensile strengths. With any luck a serious crack in a large steel structure will be spotted by an inspector and remedial action will be taken before anything dangerous happens. As we have seen 20 per cent of ships are found to have serious cracks in their hulls, say a metre or so long. If the ships had been made of a material like glass they would have broken long before a crack became conspicuous. This is basically why engineers are traditionally addicted to 'ductile' materials and really one cannot blame them.

However, in any material, a crack may sooner or later reach its critical length, whether this be a micron or ten metres. Once this has happened, as the crack gets longer, everything gets worse. The stress concentration gets worse and the Griffith energy balance gets more and more favourable to crack propagation. If the load is maintained, the crack therefore accelerates rapidly and soon reaches its theoretical calculated speed which is generally about 38 per cent of the speed of sound in the material. In the case of glass this crack speed is 4,000 miles an hour or 1,700 m/s, a velocity which has been confirmed experimentally. By this time stress waves are probably racing about in the material in all directions at the speed of sound (that is faster than the crack), being reflected off both old and new surfaces, and we are likely to end up with not one crack but with a great many. In other words the material has shattered. This is possible because, at high stresses the total strain energy in the material will 'pay' for a great many new surfaces, indeed at the theoretical strength it will 'pay' for dividing the whole material up into individual sheets of molecules.

A completely brittle material like glass is reasonably safe as long as we are content to operate it at a very low stress level, for instance as a shop window, because the Griffith crack length is then quite long and so the material is safe against minor chips and abrasions. If we want to work at a high level of stress, however, anywhere near the potential strength of glass for instance, we must be prepared to keep the surface free from even the most microscopic cracks for, if even one crack is allowed to exceed the Griffith length, which may be only about a thousand Ångströms, catastrophic failure will occur. This makes homogeneous brittle materials much too dangerous to use in practice when there is any serious stress on them.

The ability to propagate cracks freely under a small stress was not wholly a disadvantage to primitive man who was thus able to shape flint and obsidian, which are more or less natural glasses, into various cutting tools. When the flaking is done skilfully, only light hand pressure with a piece of wood is needed to detach long slivers of material which can themselves be used as knives. Using a non-brittle stone, such as jade, shaping could only be carried out by the infinitely more laborious process of grinding. Most of the tensile stresses in tools are due to bending and by keeping stone tools short and compact stresses could be kept low enough to ensure reasonable durability. A weapon like a stone sword would of course have been quite impracticable.

TOUGHNESS IN NON-METALS

The history of the attempts to prevent cracks spreading or to evade the consequences, is almost the history of engineering. The most obvious way of preventing cracks from spreading in brittle materials is not to use them in tension, in other words to use them in compression. This is what masonry is about and as we have seen in Chapter 2, starting with the simple wall one can go from the arch to the dome and to the most complicated cathedral, keeping everything in compression, or at least trying to do so. Masonry is extraordinarily satisfactory in its way but it is inherently heavy and immobile. There are therefore a number of variants of the same idea. One of these is prestressed concrete in which the brittle component is kept in compression by strong tension wires. Then there is toughened glass, which is homogeneous in the sense that it is all glass, but the outside, which is susceptible to cracks, is put into compression at the expense of a tension in the protected middle.* This has been widely used for car windscreens and, in a much more sophisticated form, is being developed for serious engineering in America. This may turn out to be an important development in new materials. Rather curiously, this line of thought does not seem to occur at all in biological materials which all depend, as do some artificial

---

*This is usually done by chilling the outside of the hot glass by means of air jets. Glass, like other substances, contracts as it cools and if the outer parts of the glass be cooled and hardened before the inner parts then the outside will, of course, contract more initially than the inside but, since the inside is still soft, it will yield in the early stages of cooling. In the later stages of cooling however both the inside and the outside behave elastically and thus their contractions get out of step. As a result, when the glass has finished cooling the outside is in compression while the inside is in tension.

Strains in glass as in most other transparent solids — become visible in polarized light and sunlight is partially polarized especially when it is reflected from non-metallic surfaces such as a road or the paintwork of a car. In this way the pattern of the air-jets used for cooling the glass can often be seen in car windscreens. Naturally the effect becomes more pronounced when one wears polarizing glasses.

ones, on reducing the effective stress concentration at the tip of the crack. Again, the methods used by nature are quite widely different from those which the metallurgist employs.

It is rather strange that so little attenion has been given so far to the mechanical properties of biological materials, though perhaps in human terms it is understandable. Many people become biologists and doctors by a reaction against things mechanical and mathematical and contrariwise engineering has been going through a phase of rejecting natural materials. Metals are considered more 'important' than wood, which is hardly considered worthy of serious attention at all.

Cellulose, which is the main constituent of wood, cane, bamboo and all vegetable fibres, is very tough. Cricket bats are made of willow, mallets of elm, polo balls of bamboo roots and loom shuttles of persimmon, a tree which grows in Persia. Aeroplanes used to be made of wood and gliders still are; wooden ships are supposed to be more robust in ice than steel ones. Cellulose cannot be considered as either weak or brittle yet it is chemically a sugar, being made by stringing glucose molecules together. All crystalline sugars are very brittle and so are glassy sugars, that is toffee. Bones and teeth are made from quite simple inorganic compounds which in their normal crystalline and glassy forms are very brittle. Of course it is possible to break a bone or a tooth but this is comparatively rare. Teeth are especially worthy of admiration, being capable, with proper maintenance, of cracking nuts for something like forty years. Even the most modern dental cements are very much weaker and more brittle than tooth substance.

The interface as a crack-stopper

The incentive to investigate the toughness of materials of this kind came, not from the biologists, but from the development of reinforced plastics which present an interesting paradox in the matter of toughness. The commonest, or at any rate the best known, reinforced plastic is fibreglass. This material consists of a very large number of thin glass fibres glued together with a resin. Now the fibres are chemically and physically in no way different from ordinary bulk glass, which, as we have seen, is catastrophically brittle. Furthermore, the resin which is used to bond the fibres is also brittle, perhaps not quite as brittle as the glass, but very nearly so. When the two are put together however, the result is a material which is sold in large quantities primarily on account of its extreme toughness.

A few years ago, John Cook and I set out to examine this effect quantitatively. The difficulty with so many problems in materials science has been that the algebra and arithmetic, though theoretically soluble, are too laborious to be done by traditional methods. This applies to a certain extent to the solution of the detailed stress distribution around a crack and until this was known in some detail it was not possible to predict what a crack would do when it met an inhomogeneity such as the interface between a fibre and a resin.

Nowadays computers are changing all this. The stress concentration at a crack tip was first calculated by Inglis in 1913, as we have said, and his results are classic and correct as far as they go. Since then a number of people a great deal abler than ourselves have worked on the problem but, because of the sheer labour of handling the algebra they were forced either to assume that the crack tip was infinitely sharp, that is had zero radius, or else, with a finite tip radius, to use very approximate methods, or at least to map the stress system only in certain regions. The assumption of an infinitely sharp crack tip leads to infinite stresses at the tip and this is clearly meaningless when one wants to investigate fracture. The other approximation using a finite tip radius did not give sufficient detail close to the actual crack tip where fracture was occurring.

Computer or no computer, I should certainly not have been able to handle the mathematics myself but John Cook likes that sort of thing and using the Mercury computer at Farnborough he was able to map the stresses very close to the tip of a crack which had a finite tip radius.

The rough general picture is of course very much that indicated in Figure 1 of Chapter 4. Generalizing a little, we might plot the stress trajectories, that is the direction in which the stress is handed on from one atomic bond to the next, very much as in

Figure 2 of this chapter. This will perhaps serve to put John Cook's detailed maps of the tip region into perspective. He and I were of course aware that we were making two assumptions which were obviously incorrect. One was that the crack tip was elliptical or rounded, which is clearly not the case in detail in a

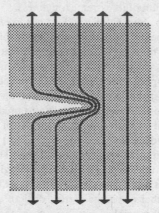

Figure 2. Rough diagram of stress trajectories in a bar uniformly loaded in tension and containing a crack. The stress trajectories become crowded together near the tip of the crack and so the stress in this region is increased.

material made out of atoms. Secondly we assumed that, elastically, the material behaved as a continuum and obeyed Hooke's law. This is also not true in detail either. However, this was the best we could do and we hope that the consequent errors are not too great.

One of the things which John Cook found in his exercises with the computer was that, as far as the stress distribution at the actual crack tip region is concerned, it does not very much matter how the load is applied. Whether the crack is forced open by a wedge, such as a nail or a chisel, or whether it is opened by a remotely applied tensile stress or bending load affects the general stress distribution in the whole body of the material very greatly, but, as far as the region sensitive to fracture is concerned, that is the area a few molecules wide in the region of the tip, the stress pattern is identical. This means that the mechanism of failure is likely to be the same by whatever means the material is broken. This was an important simplification and a step forward.

We may now turn to Figures 3 and 4, which are the actual stress distribution maps which John computed for a crack 2 microns long and 1 Ångström tip radius. The shaded area is empty space at the notional crack tip. The curved lines are not stress trajectories but are contours of stress concentration for stress at right angles and parallel to the plane of the crack. The numbers on each contour represent the stress concentration factor K by which the mean stress remote from the crack is multiplied at each point. As the crack gets longer, the tip radius remaining the same, the stress concentrations get more severe but the pattern and proportions remain similar. The reverse is true, of course, if the crack is shorter.

Figure 3 shows that the stress at right angles to the surfaces of the crack, that is the force tending to open and extend it, is very severe and is concentrated very close indeed to the crack tip. In fact the worst stress is concentrated within an area about equivalent to that covered by a single atomic bond. Its numerical value is the same as that calculated by Inglis for this critical point, though one should not attach too much importance to the exact numerical values in any of this work since all the presuppositions are approximations. However, by the time we have moved forward from the crack tip to roughly the position of the next interatomic bond, the stress has fallen to rather less than half the peak value. Such values are probably roughly true and they show fairly vividly how a great part of the load in a material is concentrated upon a single line of atomic bonds at the tip of a sharp crack, remembering of course that a material is a solid and that a crack tip is a line in a three-dimensional picture. Once the critical bond at the crack tip has broken, the peak stress concentration is transferred to the next bond, and so on, like a ladder in a silk stocking.

Merely using stronger chemical bonds will have a small effect upon the strength of a cracked material by comparison with the stress concentrating effect of the crack and this is why diamond and sapphire are brittle and usually not especially strong, in spite of their hardness and high chemical bond energies. So long as we

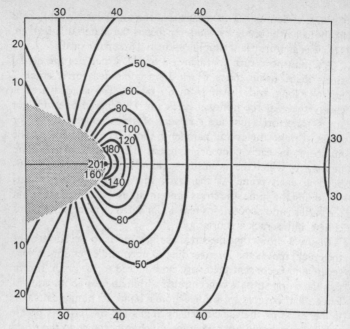

Figure 3.   Stress system close to the tip of an elliptical crack. This is a map of the stress concentrations for stresses at *right-angles* to the plane of the crack, that is, parallel to the applied load. The shaded area represents the crack tip – that is, empty space. The curved lines are contours of equal stress concentration. The figures indicate the number of times by which the local stress is increased as compared with the mean stress remote from the crack. The maximum concentration of stress is about 200. The actual values of the stress concentration will, of course, vary with the length of the crack but the relative proportions remain constant.

are dealing with materials which are approximately elastic and approximately homogeneous this is virtually the whole story as far as strength and brittleness are concerned. It does not make any appreciable difference whether the solid is glassy or crystalline or even polymeric, nor does it matter whether it has a high or a low Young's modulus so long as it more or less obeys Hooke's law, virtually up to failure. *Brittleness is not a special condition, it is the normal state of all simple non-metallic solids which cannot generate dislocations and so achieve a high work of fracture.*

Toughness requires some increase in complexity; in a sense it is a property which has to be designed for. In non-metals we can achieve an effective toughness either by devising some kind of work of fracture mechanism without making use of dislocations or else by stopping cracks from running by one means or another. Useful non-metallic materials often work by some combination of the two. (We can sometimes stop a crack by increasing its tip radius – it is fairly common to see holes drilled at the ends of cracks in glass and Perspex in the hope of preventing the crack from spreading any further.) Tough non-metallic materials often contain interfaces or planes of weakness within the materials and many of them are heterogeneous, that is to say they consist of two or more constituents, such as fibre and resin.

We must refer now to Figure 4 which is a map of the stresses *parallel* to the surfaces of the crack. On a first consideration one would not think that there would be appreciable stresses parallel to the crack surface but on reflection it will be seen that this must always be so. As Figure 2 shows, all the stresses have to get round the crack tip and are sharply bent in doing so. These stress trajectories can be thought of, more or less, as strings under tension and they will try to straighten in much the same way. If one passes a string round a post and pulls on both ends there will be a force on the post in the direction of the pull which is reacted by the post pushing in the opposite direction. In other words, there must be a tension in the area just ahead of the crack and in a direction parallel to the crack surfaces. John Cook's computations explored the distribution and magnitude of this tension and the results are mapped in Figure 4.

Unlike the stress at right angles to the crack surfaces which is mapped in Figure 3, this tension starts at zero at the crack tip and increases as we move forward away from it. The maximum is reached at one or two atomic spacings ahead of the crack but the distribution is not very peaky and a fairly high tensile stress level exists over a considerable area ahead of the crack. *Irrespective of the proportions of the crack or the means by which it is loaded the ratio of the maximum value of this stress parallel to the crack surfaces to the peak opening stress at right angles to the crack is constant and has a value of one to five.* This state of affairs seems to be fundamental to all cracks existing in a stretched material.

This is where the internal surfaces in biological materials and reinforced plastics become important. It is significant that these interfaces are generally weaker than the surrounding material. This is not because Nature is too incompetent to glue them together properly but because, properly contrived, the weak interfaces strengthen the material and make it tough.

Consider what happens when a crack approaches an interface of this kind which is roughly at right angles to it. When the area

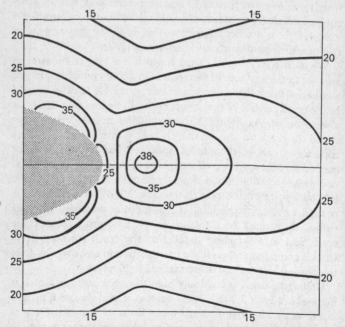

Figure 4.   Stress system close to the tip of an elliptical crack. This is a map of the stresses *parallel* to the plane of the crack, that is, at right-angles to the applied loading. For this case the maximum stress concentration is about 40, i.e. one-fifth of that in Figure 3.

of tension stresses ahead of the crack tip reaches the interface, it will try to open it by pulling the two sides apart. If the strength of the interface is greater than about one-fifth of the general cohesion of the material then the interface will not be broken, the crack will cross it and the material will behave as a normal brittle solid. *If however the adhesive strength of the interface is less than about one-fifth of the general cohesive strength of the solid then the interface will be broken before the main crack reaches it and a crack trap or crack stopper has been created.**

This is shown diagramatically in Figure 5 and in the flesh, in a reinforced material in Plate 11. Of course, if the adhesion at the

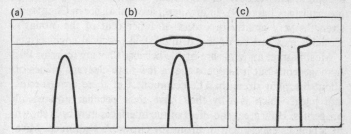

(a)   (b)   (c)

Figure 5. Cook-Gordon mechanism for stopping cracks at a weak interface.
   (a) Crack approaches a weak interface.
   (b) Interface breaks ahead of main crack.
   (c) T-shaped crack-stopper. In practice the crack is usually diverted, as in Plate 11.

interface is too weak then the material as a whole will be weakened so that, when there is no adhesion at all, one has to have some arrangement like cloth or rope or basket-work to hold the material together by friction. For the best results, exact control of the adhesion seems to be critical but when this is achieved, as it seems to be in the best natural and artificial composite materials, very excellent combinations of strength and toughness are created.

*I suppose that one could consider that the crack has been blunted. The tip radius, from being very small, is now very large, practically infinite in fact. It is true that we have now a new crack, at right angles to the original one, but then the tendency to propagate a crack which is *parallel* to the applied stress is usually nil.

Since the condition for effective crack-stopping is that we have to weaken the material by a factor of five, the process does not sound a very promising one. We have, as it were, given up before we have started. As toughening processes go, however, the interfacial weakness method seems to be rather efficient. This is because the true fracture stress at the crack tip is presumably the theoretical strength of the material and this generally lies between 10 and 20 per cent of the Young's modulus, E (Chapter 3). Reducing this to one fifth still leaves a potential strength of between 2 and 4 per cent of E, which is roughly what is achieved in practice in fibre-glass and is a good deal more than one can reach, keeping a safe amount of toughness, with the metallic ductility mechanism (Chapter 9). In any case, as we shall see in Chapter 8, strengths very greatly in excess of 1 per cent of the Young's modulus may not be of much interest in practical engineering.

Most natural minerals are brittle because they are more or less homogeneous but it happens that a few have cleavage planes of about the right strength. The commonest of these are asbestos and mica, which is why they have their peculiar and useful properties. How great the effect of the interfaces may be is shown by a famous experiment of Professor Orowan's with mica. Mica is an ionically bonded mineral in which, because of the arithmetic of the electrical charges in the molecule, every so often there is a layer of metal atoms in the crystal in which each atom has to share a single electron's worth of charge with several near neighbours, so that this layer in the crystal is a weak one. The useful form of mica is called Muscovite (because it originally came from Russia) and in this mica the strength of the bonds on the weak cleavage plane is nominally and very roughly a sixth of the strength of the bonds elsewhere in the crystal.

What Orowan did was to measure the strength of Muscovite mica in tension. In his first experiment he cut from a sheet of mica a normal hour-glass shaped test-piece (Figure 6(b)). This test-piece was flat and quite thin and the planes of weakness lay parallel to the broad flat surfaces. The whole specimen might be regarded, on a molecular scale, as being cut from a number of sheets of paper, weakly glued together. The edges of the specimen, which had been cut mechanically, were, in detail, quite rough. When the specimen was loaded in a testing machine the edges

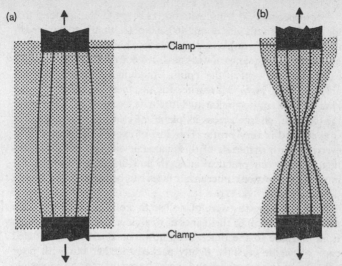

Figure 6. Orowan's experiment with mica. Effect of specimens with stressed and unstressed edges.

(a) Edges unstressed. Tensile strength 460,000 p.s.i. (3,100 MN/m²).

(b) Edges exposed to average tensile stress. Tensile strength 25,000 p.s.i. (170 MN/m²).

were stressed as much as the middle and so cracks started at the edges and spread inwards across the material in the usual way. The tensile stress developed in this test was about 25,000 p.s.i. (170 MN/m²), much the same as ordinary glass and perhaps a little less than commercial steel.

Orowan now tested a differently shaped specimen of the same mica. In this case the sheet of mica, though otherwise similar, was not waisted but was of a rectangular, playing-card shape, somewhat wider than the clamps which gripped it. It was assumed that the stress followed a path similar to that sketched in Figure 6(a). Thus the edges were largely unstressed. The outside surfaces lying on the stress-path between the grips were, of course, fully stressed and no doubt contained all manner of iniquities in the way of scratches and stress concentrations. For these to extend, however, the crack would have had to cross the planes of weakness in the crystal which were in their path.

N.S.S.M.–6

For a specimen of Muscovite of this shape Orowan found that the tensile strength was about 460,000 p.s.i., that is to say nearly twenty times as strong as a specimen in which the cracks did not have to cross the planes of weakness. 460,000 p.s.i. (3,100 MN/m²) is about 1½ per cent of the Young's modulus and a very respectable strength. Now Margarite, for instance, which is another kind of mica, quite similar to Muscovite except that it has twice the electrical charge across its planes of cleavage, has negligible strength and is very brittle. This sort of experiment shows however that with materials of this character one cannot really distinguish between practical strength and brittleness so that the introduction of weak internal surfaces can be regarded as raising the strength.

Mica and asbestos were of no use to stone-age men for tools and weapons because the planes of weakness run straight through from one side to the other. Jade however consists of a tangled mass of needle crystals, tightly packed together but with poor adhesion at the interfaces and might be regarded as an inorganic equivalent to a briar pipe or a bamboo root. Jade is therefore very tough and would have been almost ideal for tools and weapons if only it had not been so difficult to work and so scarce.

Since jade cannot be flaked, like flint and obsidian, it could only be shaped by grinding it with sand on a piece of wood for weeks or months. Hence, though very durable, jade implements were very costly and partly for this reason, and partly for the beauty and scarcity of the material itself, they remained symbols of prestige after the introduction of metals.

Jade is scarce because it can only crystallize in that form under geological conditions of heat and pressure which were confined to certain faults in the earth's crust. These occurred in the Far East, in New Zealand and in Central America. Jade axes were being made by the Maoris in New Zealand, almost within living memory. According to Heinrich Harrer axes are being made from a jade-like stone in central New Guinea today. He says that the polishing takes several months.* A curious problem is set by the discovery of a few jade axes in England recently. If these are not the product of a Piltdown-type hoax, then either there must have

* Heinrich Harrer, *I come from the Stone Age*, Hart-Davis, 1964.

been a source of jade somewhere in Europe or else the axes must have been brought an unimaginable distance from the Far East, a journey comparable in its way to that of the monoliths of Stonehenge. However, as Herodotus remarked on finding Scythian artifacts in Delos, they may have 'diffused'.

The instances of effective crack-stoppers in minerals are fortuitous. When one looks at biological materials one is impressed with the enormous care which Nature seems to take over the interfaces when she is being, as it were, teleological. A good example is the construction of teeth, about which a certain amount is known. Teeth consist of a hard, tough surface layer called enamel while the interior is made of a material called dentine. Both constituents however contain elongated inorganic crystals distributed in an organic matrix and the principal difference between enamel and dentine lies in the proportion of inorganic material to organic material.

The hard part of both enamel and dentine consists of elongated crystals of a substance which is nominally hydroxyapatite, $Ca_{10}(PO_4)_6(OH)_2$, although the exact chemical composition varies widely, reflecting the environment in which it was formed, and carbonoapatite, fluorapatite, calcium fluoride, calcium carbonate and so on may be present. These crystals are quite small and in enamel are about 3,000 to 5,000 Å in length and from 500 to 1,200 Å thick. In enamel these crystals are very closely and beautifully packed together so as to constitute 99 per cent by volume of the material. They are separated by a thin layer of a very complex organic material. This separating layer is undoubtedly mainly a protein and was originally thought to be similar to keratin which is the protein in hair. It is now thought that it is a special protein unique to tooth enamel. Incidentally it changes its composition considerably between the foetus and the adult.

Dentine differs from enamel in that the inorganic part constitutes only about 70 per cent by volume. The apatite crystals are also much smaller being 200–300 Å long and 40–70 Å wide. These apatite crystals are embedded in an organic matrix which is mostly collagen.

The adhesion between the hydroxyapatite and the separating

protein layer is of a most complicated chemical nature being partly by hydroxyl bonds and partly by ionic bonds (see Appendix 1). No doubt very precise control is exercised over the interfacial adhesion and thus over the propagation of cracks. The presence of the weak organic layer, however, must enable decay to get a start and, once the enamel is penetrated, the high organic content of the dentine enables the attack to propagate fast. Apparently one cannot have it both ways. If the vulnerable organic layers did not exist teeth would not rot so easily but then they would be brittle and would probably break early in life.*

The use of hydrogen bonds, that is hydroxyl groups (−OH) (see Appendix 1) to control adhesion at an interface is very common in living organisms and it is presumably a convenient method where the environment is continuously wet. When natural materials are used by man in a dry environment difficulties arise. The drying of the hydroxyls, each of which normally has a shell of water molecules around it, leads to the shrinkage of materials like wood. It can also lead to drastic embrittlement, because the interfacial adhesion gets too strong. This can be the case with ivory, which is pretty nearly tooth material. The Parthenon in Athens contained a famous gold and ivory statue of Athene and the Parthenon must have been even hotter when the roof was on than it is today. To preserve the ivory from becoming brittle and cracking, the statue was surrounded by a shallow pool of water which, besides reflecting the light upwards on to the statue, maintained the humidity high in the naos. This pool was kept topped up, and the statue preserved, for about eight hundred years. The remains of the rim of this pool, which was only an inch or two deep, can be seen on the floor of the Parthenon today.

Thermoplastics

So far we have dealt mostly with metals and with natural non-metallic materials of various kinds. The large class of artificial non-metallic solids which we generally call 'plastics' are all based on arrangements of long-chain molecules, usually made preponderantly from carbon atoms. The earliest material of this sort

* I am indebted to Mr J. W. McLean for this account of tooth structure

was 'Bakelite', a polymer made by reacting phenol with formaldehyde; it was patented in 1906. Such materials have the advantage that they can be made permanently hard by heating them to a temperature around 150°C, for which reason polymers with this sort of constitution are called 'thermosetting'. However, because thermosetting plastics consist essentially of an irregular three-dimensional network of molecules, something like glass, they have quite a low work of fracture – seldom more than 100 J/m² and usually much less – and so they behave mechanically in much the same way as glass does. As we shall see in Chapter 8, the only way to make thermosetting plastics reasonably tough is to incorporate fibres of one kind or another, in other words to use them in a composite material. Although the addition of fibres is a fairly effective way of making these materials tough it does put up the cost of manufacture and it also restricts the number of applications for which they are suitable.

Many of the objections to plastics of the thermosetting type for household goods and such-like applications have been got over by the development of the 'thermoplastic' resins which came into use on a large scale after 1945. These materials are based on polymeric chains which are not cross-linked, in other words the molecular arrangement is quite different from that of glass. As the name implies they do not harden permanently but will soften repeatedly if one subjects them to some temperature between about 100° and 150°C. Although this characteristic is to some extent a handicap in that they cannot be used, for instance, with boiling water, there are very great manufacturing advantages since the stuff can be squirted out hot in the form of tubes and sections or else into an elaborately shaped mould where it can be hardened almost instantaneously by rapid chilling. These processes are known as 'extrusion' and 'injection moulding' and for many purposes they have proved the cheapest of all manufacturing methods.

Nowadays such materials exist in a great many chemical varieties and their trade names are legion though 'Polythene' and 'Nylon', together with 'P.V.C.' (polyvinyl chloride) are probably still the most popular and the best known, at least in this country. The really phenomenal success of these materials during the last thirty years or so has been due to the combination of cheap and rapid mass-production with adequate toughness – added of course to chemical inertness, lightweight and bright and cheerful, not to say garish, colours.

The text-book example of this sort of thing is polyethylene (Polythene). This consists, basically, of $(CH_2)_n$, in other words it is a linear long-chain molecule much like Figure 7.

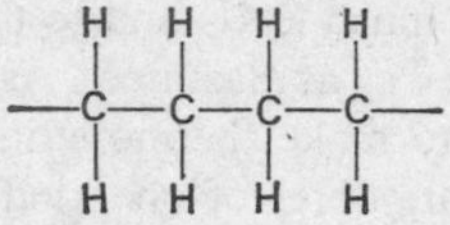

Figure 7.

Other synthetic polymers are chemically rather more complicated but, elastically, apparently not very widely different. But now consider that the E of diamond – which has similar carbon-carbon bonds – is around 170×10^6 p.s.i. (1,200,000 MN/m²). Diamond has a density of about 3·5 grams per c.c., polyethylene about 0·92; when we have made all the necessary corrections however we find that we ought to expect an E of about 10×10^6 p.s.i. (70,000 MN/m²) for polyethylene and similar thermoplastics.

In experimental fact, the E for polyethylene is around $0·3 \times 10^6$ p.s.i. or 2,000 MN/m² – about a thirtieth of what theory predicts – and most of the other synthetic polymers are not much stiffer. What is remarkable about the text-books (which describe the chemical structure of the various chains in loving detail) is not so much that they do not explain the discrepancy but that they do not even notice it; nor are they apparently interested in the fact that the work of fracture for a material like Polythene or Nylon is at least a hundred times higher than it is for most of the thermosetting plastics.

Owing to the work of Frank and Keller at Bristol University the probable explanation is now apparent. It has long been known that the chain molecules of many natural and artificial polymers are arranged in a way which is at least partially crystalline. What emerges from Frank and Keller's work is that the nature of the crystals is different in natural and synthetic polymers. As we shall see in the next chapter, in natural materials like wood, the long-chain molecules are arranged roughly parallel to the length of the tree, that is to say, more or less in the direction of the most important stresses. This is why timber develops a good fraction of its theoretical modulus in actual practice.

In materials such as polyethylene, however, Keller finds that the long $(CH_2)_n$ chains are arranged in quite a different sort of way. In fact they are folded on themselves in a zig-zag pattern something like Figure 8, for polyethylene the fold length is usually pretty constant at about 180°Å.

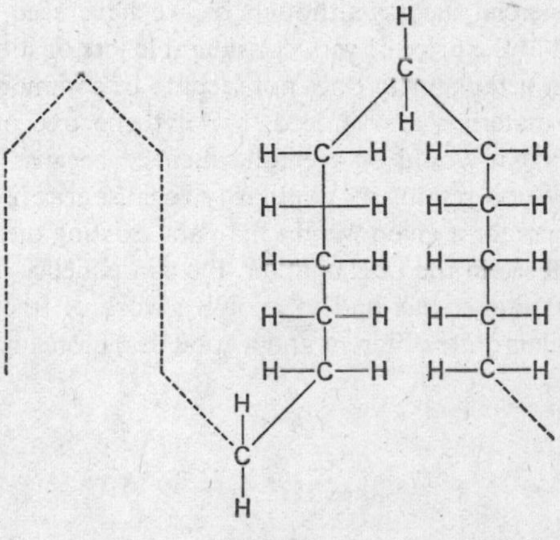

Figure 8. The long flexible chains of polymers like polyethylene frequently crystallize in a folded or zig-zag pattern.

Examining this structure with the eye of an engineer it is at once evident that very little of the stiffness of the carbon – carbon chain will be reflected in the macroscopic modulus of the plastic since the bonds which control the extension are not the covalent primary bonds but the secondary or van der Waal forces which attach the convolutions of the chain to each other.

The experimental values for the E of polyethylene do, in fact, fit very well with both the known stiffness of the van der Waal forces and also with that of molecular crystals of the phthalo-cyanine types. (Chapter 2)

Thus the folded, Keller-type crystallization results in a large reduction in the stiffness of thermoplastics of the polyethylene type. This may, or may not, be a technological handicap, depending upon what the material is meant to do. It does seem, however, that Keller crystallization is wholly beneficial in regard to the work of fracture of these materials. Before the plastic can break, many of the crystals have to unfold, rather like concertinas, and, in doing so, a great deal of energy is expended in unravelling the long molecules (Plate 13).

It is to this property that thermoplastics owe a good deal of their commercial success although, as we have seen, toughness is obtained at the price of very considerable loss of stiffness. Such a molecular arrangement does not seem to be common in natural structural materials and indeed, a Polythene tree of any considerable height would be a public menace because of its low modulus. Wood retains its popularity because it can offer much more stiffness for a given weight than any existing thermoplastic. As we shall see in the next chapter, the consequence of this stiffness is that timber has had to evolve a work of fracture mechanism which is quite different and a good deal more ingenious.

or Wooden ships and Iron men

'*Plastics are made by fools like me
But only God can make a tree.*'

During the war, when we were doing research on strong plastics, Professor Charles Gurney used to recite this little ditty to me nearly every day and I found it depressing because wood was in fact a better material for making aeroplanes than the plastics which we could then produce. Even today there are classes of structures such as sailplanes and some kinds of boats for which wood is still the most efficient material available.

Not only are wood and other forms of cellulose technically efficient but they are also fantastically successful, judged by any quantitative criterion. Cellulose is the structural part of all vegetable matter and it is the strength and stiffness of cellulose which displays leaves and greenery to the sunshine so that photosynthesis can take place and become the principal chemical starting point for all forms of life. Cellulose forms on the average about a third of the weight of all vegetation and the world tonnage of plants is almost beyond computation, locking up in cellulose a large fraction of the world's limited supply of carbon. Cellulose seldom occurs in animals but there is one rather dim little class of marine animals, the Tunicates, which are mostly made from cellulose. They look rather like elongated jellyfish and appear to have no structural virtues. However chitin, the structural polymer in insects, is very similar to cellulose.

When we come to the works of man, cellulose is still in the leading place. If we consider the timber which is sufficiently industrialized to get into the official statistics, the annual world consumption (not counting fuel) appears to lie between 800 and 1,000 million tons. The rough timber, fencing, bamboo, reeds, thatch and so on used by farmers and primitive people may possibly amount to nearly as much again, but naturally, no records are available. The world production of iron and steel is somewhere round 450 million tons, that of all other metals is negligible by comparison.* Since, weight for weight, the strengths of commercial steel and timber are comparable, the total of the burdens supported by wood may well be greater than those supported by steel, though no doubt many of the loads which steel carries are the more spectacular.

Since the density of wood averages about one-fourteenth of that of steel it may be that about thirty times the volume of wood is used, taking the world as a whole.

The ratio of the consumption of wood to steel varies considerably between different countries but it is not necessarily an index of the degree of industrialization or of technological advancement. England and Holland both use about 1,100 lb. of steel per head per annum as against about 700 lb. weight of wood. In U.S.A. the consumption of steel per head is about the same, 1,100 lb., but the consumption of timber per head is much more, about 2,400 lb. In Canada it is as high as 3,300 lb. per head per annum. The characteristic of less developed countries is that their consumption of both wood and steel, that is the total tonnage of their artifacts, is less.

Plant growth

Cellulose is an example of standardized production on the part of nature. Although plants vary so greatly in their shape, function and general appearance, the cellulose molecule is the same in all. It may vary slightly in length and in its physical arrangement but these are matters of detail; the chemistry is the same.

All the more advanced plants contain hollow, elongated, spindle-shaped cells (Plate 13) whose walls are made largely of cellulose. (Which is why it is called 'cellulose', '-ose' being the chemical termination for sugars, 'fructose' is the sugar found in

---

*It is rather difficult to get a comparison in money values. The relative prices of timber and steel vary greatly in different countries and also the price of timber itself varies from that of rough timber, which may be much cheaper than steel, up to expensive plywoods which cost far more than steel sheet. Again it depends at what stage of manufacture one makes the comparison. Very, very roughly, however, the cost of 'industrial' timber is about the same as that of commercial mild steel per lb.

fruit, and so on.) These hollow spindles are the fibres which take the loads and provide the strength.

Initially the simple sugar, glucose (Figure 1), is synthesized in

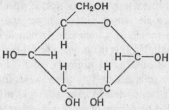

Figure 1. The glucose molecule.

leaves from atmospheric CO_2 and water by the action of sunlight on the green catalyst chlorophyll. Like other simple sugars, glucose is soluble in water (which is why it is easily digestible) because it has five hydroxyl groups (see Appendix 1) which have a strong attraction for water molecules, and also because the glucose molecules are physically small enough to shuffle around fairly freely in a liquid, provided there are not too many of them. Concentrated solutions of glucose approximate to treacle.

Glucose in dilute solution in sap-water thus passes through internal passages in the plant until it reaches a growing cell. In the wall of the growing cell the glucose molecules are joined together

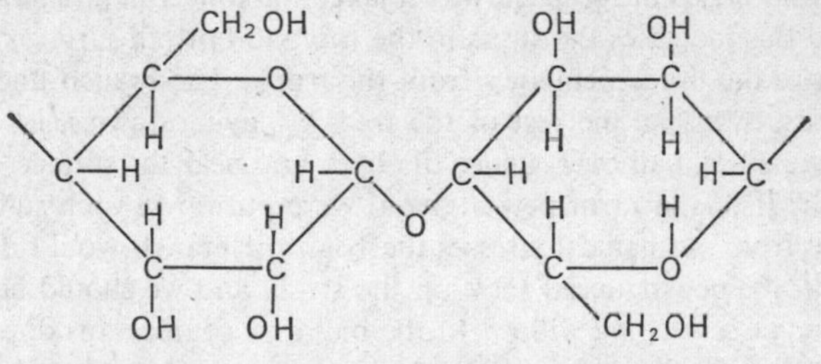

Figure 2. The cellulose chain. It is usually several hundred glucose units long.

endwise (Figure 2) by a chemical reaction known as a 'condensation reaction':

$$-OH + HO - \rightarrow -O- + H_2O$$

The result is an oxygen linkage and a molecule of water which goes off in the sap.

This process is controlled in the plant by substances called 'auxins' though how it is done is not at all clear. The oxygen linkages between the sugar rings remain the vulnerable links in the cellulose molecule which may reach a length of several hundred glucose units. It is the oxygen link which is broken by the enzymes in the stomachs of animals, such as sheep and cows, which can digest cellulose and by the various fungi or rots which attack wood. It is also the linkage which is attacked by simple chemicals, such as bleaching powder, which are used by laundries, and accounts for the gradual weakening of shirts in the wash.

The cellulose chains which are laid down in the cell-wall are long and they have their length more or less parallel to the length of the cell or fibre, that is to say in the direction of the applied stress. The growth of cellulose is altogether a very remarkable business. If we consider an ordinary tree, by the time it is a few years old it has usually acquired a number of little branches, coming out more or less horizontally from the main stem or trunk. Each of these little branches is in effect a cantilever beam stressed in bending by its own weight (Chapter 2). This means, as we have seen, that the upper surface of the branch is stressed in tension and the lower surface in compression, like any other cantilever. As the bough grows thicker and longer, it gets heavier, and this increases the stress in the top and bottom surfaces near where the branch emerges from the trunk. The branch thickens and grows, like the rest of the tree, by laying down a layer of new material all over, under the bark and near the surface, each year. If this layer of new material were put down each summer free from mechanical stresses the beam or branch would droop until the new material took up the strain and we should have a tree like a weeping willow. In the majority of trees this does not happen. The boughs grow out from the trunk at nearly the same angle throughout the life of the tree and the sapling can be regarded as a geometrical model of the fully grown tree. It follows that, in the majority of trees, the new cellulose is laid down in the cell already containing the stresses and strains which it has to bear.

Working with hydroquinone and other fairly simple soluble substances, I have grown long needle crystals or whiskers (Chapter 4) which thicken by the growth of sleeve-like surface layers which are geometrically not unlike the growth layers in a tree. The initial whisker crystal or filament is often highly bent and the growth layers can be seen to exert a very strong straightening

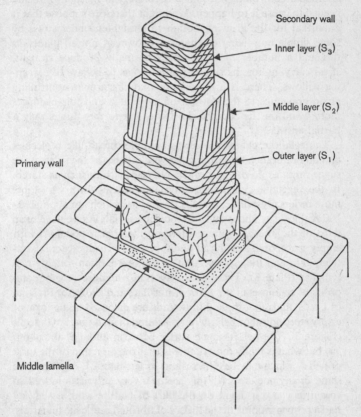

Figure 3. The cells in timber are very roughly rectangular in cross-section. The morphology of the cell walls is complicated but the disposition of the cellulose molecules and fibrillae is preponderantly helical, something like this diagram. (© Crown copyright.)

action on the bent filament, such that, by the time the sinuous initial thread has grown to a millimetre or so thick, it is invariably straight. From this it is clear that the growth layers of these crystals are formed under considerable mechanical stress if this is needed to straighten the crystal. This occurs quite frequently in simple, non-biological systems and there is no question of any additional controlling substance or biological mechanism being needed to cause it to happen. We might therefore suppose that it is normal for the growing bough to straighten under stress by some simple non-living mechanism. However, not all plants do this and a number of trees which normally produce straight, stress-carrying boughs can be grafted so as to behave like weeping willows. There is a suggestion that the growth-controlling auxin gravitates to the bottom of the bough and this produces more wood on the compression face, but to me, this is only a partial answer.

The cellulose chains are always simple thread-like molecules and do not branch by forming oxygen linkages at the sides of the sugar rings, as do other, weaker, polysaccharides such as starch. In the vegetable cell these cellulose molecules form very long, more or less crystalline, threads or fibrillae which are about 150–200 Å thick, say about 30 to 40 molecules wide. As we have seen, most of the internal volume of wood material is taken up with empty space, or at least by air and sap. The cell walls are comparatively thin and the cross-section of the cell is often roughly rectangular (Plate 8). These relatively thin walls are largely composed of cellulose, in the form of fibrillae, and Professor Preston, of Leeds, finds that these thin threads are disposed in the form of a very steep spiral or helix, wound around the long axes of the cells (Figure 3). The helical angle varies between about 6° and about 30° but what is really remarkable is that the direction of the twist or helix – which may be either right- or left-handed – is always the same in any one tree. All this seems a very curious – indeed an eccentric – arrangement on the part of Nature who, if she had had a proper training in the theory of fibrous composite materials, would surely have known better.

A little while ago Dr Giorgio Jeronimidis came to work with me on just this subject. The first thing that George found was that the work of fracture of wood was quite exceptionally high, although of course it has nothing resembling a dislocation mechanism to help it. In fact the work of fracture, W, turned out to be around 10^4 J/m^2, which, weight for weight, is at least as good as a ductile steel and a good deal better than 'tough' composites like fibreglass. In fact the figure is much better than one would predict from composite theory (Chapter 8), supposing wood to behave like an artificial composite. As Punch would have said many years ago, 'collapse of Stout Party'.

Since this high work of fracture – which makes trees able to stand up to the buffetings of life and which makes wood such a useful material – cannot be accounted for by any of the recognized work of fracture mechanisms which operate in man-made composites, George set out to find out what was really happening.

Now the various cells in wood are glued to each other (by means of the various non-cellulosic constituents which exist in timber) in a way which is reasonably effective, but not very effective. This is why the paper-maker is able to make a fibrous pulp for your morning paper from wood. If a crack begins to penetrate into the wood across the grain, the Cook–Gordon mechanism – which we discussed in the last chapter – comes into operation in the region around the crack tip and the various cells become separated so that each of them operates as an independent helix, something like a drinking straw. When this happens the thin walls of the tubes are able to buckle, the helical fibrillae can then straighten themselves out and so the cell is enabled to elongate under the tensile load by something like 20 per cent. Both by calculation and by experiments with model cells, George was able to show that the process of buckling and elongation absorbed a great deal of energy. This process shows great cunning on the part of Nature, also a good deal of cleverness on the part of George. It adds, very usefully, to our repertoire of work of fracture mechanisms and, as we shall see, it seems likely to turn out extremely useful in the design and manufacture of artificial composite materials.

In fact the arrangement of the cellulose molecules in wood is partially crystalline and partially amorphous. The crystalline regions are held together sideways by hydroxyls which have got rid of all their attached water molecules and once the system has locked solid into a regular crystal, the interstices of the crystal become inaccessible to water. We know that this is so because the X-ray diffraction pattern, which shows the crystal lattice spacing, does not change when cellulose swells in water. On the other hand, cellulose absorbs both liquid and atmospheric moisture very actively and this, from the engineer's point of view, is one of its worst vices.

The proportion of crystalline material in natural cellulose varies a good deal but may be about thirty or forty per cent of the whole. The non-crystalline, that is the amorphous cellulose, has no mechanism for protecting its hydroxyls from moisture, since most of them are not firmly attached to their neighbours, and so they pick up a shell, round each hydroxyl, of any water molecules which are available. This naturally reduces their attraction for each other and so the forces holding the cell wall together laterally are diminished and the cell swells. It is stopped from passing completely into solution, partly by the large size of the cellulose molecules and, more, by the fact that, in natural cellulose, the whole system is tied together mechanically by the presence of the crystals, which are water-proof and form a good proportion of the whole mass. So-called 'regenerated celluloses', such as Cellophane, are made by dissolving natural cellulose by chemical methods which break up the crystals. The resulting solution is then precipitated to form a transparent film which is largely a tangled-up felt of individual molecules, and is much less crystalline. When such films are wetted they become very flabby indeed and lose all their strength. The Cellophane which is made for wrapping and packaging is therefore protected by a very thin coating, on each face, of a water-resistant lacquer. This gives sufficient protection for its ephemeral purpose but, after prolonged wetting, such materials are hopelessly weak, whereas natural celluloses retain a good part of their strength.

The natural celluloses which we use include a very large number of timbers, bamboo, cane, flax, hemp, cotton, ramie, sisal, esparto and so on. However, as we might expect, their mechanical behaviour and especially their swelling in water and the relation between their temperature and moisture content and their strength, differ only in detail and present much the same general picture.

The properties of wood

Trees grow in all shapes and sizes and their timbers look very different. These variations are however more or less superficial and the main differences between timbers lie in their density. Seasoned balsa has density of five to ten pounds per cubic foot (s.g. 0·1), spruce around thirty (0·45), oak about fifty (0·7) and lignum vitae between seventy and eighty (1·1). With quite minor additions and subtractions the actual wood substance has in all cases about the same chemical constitution and about the same density of ninety pounds per cubic foot (that is, much the same as sugar – say 1·5).

As we have said, the main structure of wood consists of large numbers of tubular cells or fibres of squarish cross-section fitting very neatly together (Figure 3 and Plate 8). There are minor distinctions in the geometrical arrangement of the fibres in different species. For instance, some timbers, notably oak, have a certain number of fibres, medullary rays, running radially in the trunk and thus crossing the longitudinal fibres at right angles. From the engineering point of view, however, all woods may be considered as bundles of parallel tubes, rather like bundles of drinking straws. Since the tubes are made of substantially the same material the large range of density is caused by the various thicknesses of the cell walls. One consequence of this is that, to a first approximation, most of the mechanical properties of different timbers are proportionate to their densities; a timber twice as dense will be about twice as strong and so on. This is not quite true but it is roughly so.

Wood substance consists of about sixty per cent of cellulose, various other sugar compounds and lignin, a substance having affinities to a resin, which impregnates adult wood substance in some fairly intimate way. Unlignified cellulose is birefringent, that is to say, it rotates polarized light because of its highly directional nature and it also stains brightly with certain dyes. Normal wood substance containing lignin does not do either of these things.

However, immediately before mechanical failure, and before any weakness can be distinguished by mechanical methods, wood becomes both birefringent and easily stained by characteristic dyes. This is probably due to the early stages of George Jeronimidis' fracture mechanism of which we talked on page 135. These phenomena cannot be used as a warning of incipient fracture because, to observe the effects, it is necessary to cut thin sections of the stressed part and to look at it in an optical microscope. However, the method can be very useful when investigating accidents and it also serves to show how subtle is the nature of wood substance. Some tropical woods such as teak and greenheart contain small amounts of toxic chemicals and also of silica. These protect the timber from insects and rots but they also help to account for the high cost of working the best tropical woods because the silica blunts tools very quickly and the splinters of greenheart are poisonous.

As we have seen, wood depends for its defences against crack propagation partly upon Jeronimidis' work of fracture contrivance – which ensures that the critical Griffith crack length is a long one – and also, by way of a further safety device, upon the Cook–Gordon mechanism for stopping any crack which gets past George. The other mechanical properties of wood are very much what we should expect from a bundle of tubes or fibres. Laterally, that is across the grain, they separate or crush quite easily, so that the lateral tensile and compressive strengths are very low, only a few hundred pounds per square inch. The lighter woods, such as balsa, can be crushed with the finger. On the other hand, it is just because the fibre tubes can be crushed locally that wood can be nailed and screwed without splitting, provided we do not abuse the wood too much. Incidentally, nails and screws of reasonable size, put in with reasonable care, do not weaken the wood, as a whole, in any measurable way,* in other words wood is astonishingly resistant to stress concentrations.

The tensile strength of spruce, for instance, is around 17,000 p.s.i. or 120 MN/m² when carefully measured. This represents an elastic strain or interatomic separation of about 1·0 per cent, per-

---

* Dr Richard Chaplin finds that, if you take a screw *out*, the wood will be weakened in compression by the hole which is left – so leave them in.

haps between a tenth and a twentieth of the theoretical strength. These figures are much better than those for most other engineering materials, especially cheap ones. A commercial mild steel strains elastically about 0·15 per cent. Weight for weight, the tensile strength of wood is equivalent to that of a 300,000 p.s.i. steel, which is four or five times the strength of the steels in common use. In practice, as we shall see, it is not very easy to make effective use of the high tensile strength of timber.

The weakness of wood is in compression along the grain. In this respect it is the opposite of cast iron, which is strong in compression and weak in tension. Again, a bundle of drinking straws glued together provides a realistic model. Under a compressive load the thin wall of one of the tubes decides to buckle or corrugate and all the rest have to follow it (Figure 4). The com-

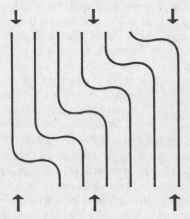

Figure 4. Compression failure in wood. On the clean, planed side-grain of the timber the failure can be seen with the naked eye as 'creases' running across the grain direction.

pressive strength of spruce generally lies between 4,000 and 5,000 p.s.i., say 30 MN/m². Weight for weight, this is still quite respectable, as compared with steel, but it is of course much less than the tensile strength.

When wood begins to fail in compression little lines of buckled fibres can just be seen running diagonally or across the grain but these are easily missed unless the surface is clean and you know what to look for. For some time after the initial failure nothing very sensational or catastrophic happens, the wood just yields gradually. In most cases wood is used in bending and the result of gradual crushing on the compression side of a beam is to transfer load to the tension side. In this way, the nominal stress in a wooden beam before actual collapse occurs may be up to twice the true compressive stress. It is this which makes a structure made out of timber such a safe one, generally one can very nearly get away with murder. Again, timber is noisy stuff and it will frighten the wits out of you before it is in any real danger of breaking. Sailplanes are often launched by means of half a mile or so of wire, reeled in by a winch. Having no engine, gliders are delightfully silent, except for a slight noise from the wind so that one can hear the structure very well. On a fast, gusty launch a wooden glider will treat you to a series of creaks and groans, and occasionally bangs, which are alarming until you realize that it is all pretence and that the structure is not in the least danger of breaking up. In fact it puts on this performance several times a day. I am pretty sure that these noises do not proceed from incipient compression failures. I have often wondered where they do come from but confess that I have absolutely no idea. They are however counted to wood for righteousness: as long as one can hear a timber structure one is very unlikely to break it.

For its weight, therefore, the strength of timber is as good or better than most of its competitors. Strength however is not enough: one must also have adequate stiffness. Substances like Nylon have plenty of strength but they are not sufficiently stiff to make engineering structures. The Young's modulus of spruce is about 1·5 to 2·0 × 10⁶ p.s.i. (12,000 MN/m²) and the other timbers are, roughly, more or less stiff than this in proportion to their densities. Curiously, weight for weight, the Young's modulus of timbers is almost exactly the same as steel and aluminium and much better than synthetic resins. The good stiffness, combined with low density, means that wood is very efficient in beams and columns. Furniture, floors and bookshelves are usually best made in wood and so are things like flagstaffs and yachts' masts. The railways in America could be built very quickly and cheaply in the nineteenth century partly because of the efficiency of the timber trestle bridge.

As against these virtues, timber creeps. That is to say, if a stress is left on for a long time, wood will gradually run away from the load. This can be seen in the roof-tree of an old house or barn, which is generally concave. The creep of the wood is the reason why one must not leave a wooden bow or a violin tightly strung. The cause of the creep is most probably simply that, in the amorphous part of the cellulose, the rather badly stuck hydroxyls take advantage of changes in moisture and temperature to shuffle away from their responsibilities. It is unlikely that the crystalline part of cellulose creeps to any measurable extent.

Swellulose

No doubt it would not be beyond the wit of nature to join up the cellulose molecules sideways with primary chemical bonds so that it would be thoroughly tied together and would have much the same strength in every direction. However, as we said in the last chapter, it seems to be a condition for the strength and toughness of materials of this type that there should be planes of weakness parallel to the strongest direction. If not, wood would be something like a lump of sugar: homogeneous but weak and brittle. For its weight, there is really nothing wrong with the mechanical properties of wood and the weight of wooden structures is generally at least comparable to that of metal ones. We pay for this however in the vulnerability of wood to moisture.

Wood is affected by liquid water in the form of rain, rivers, seas and so on with which it may come into contact but, more importantly, it is affected by the moisture vapour which is always present in the air.

Air at any given temperature can hold so much moisture; any excess is precipitated as rain, fog, mist or dew. Such air is called saturated and thus the 'relative humidity' on a wet day is around 100 per cent. Indoors, or in drier weather, the relative humidity decreases, although it seldom falls much below 30 per cent even in hot dry climates.

All timbers tend to come to an equilibrium with the relative humidity of the surrounding air. Exposed for a long time to moist, saturated, air timber might settle down to a moisture content of 22 per cent or 23 per cent. In a very dry climate the moisture content might reach as low a figure as 5 per cent. Regarded as mere changes of weight these figures are of secondary importance. What is important is the effect of the moisture on the wood. The most important effect is that the wood shrinks or swells. The movement in the direction along the grain of the wood is negligible, as one would expect from the molecular structure. The cross-grain swelling and shrinkage is however very large. Every one per cent change of moisture content may cause about a half per cent shrinkage or swelling. Over the range of moisture contents likely to be reached in air the lateral dimensions of wood can thus change between five and ten per cent, that is up to an inch on a ten-inch-wide plank. Amateurs rather like to use wide planks if they get the chance, professionals are wiser and prefer narrow ones so that the movement at each individual joint is smaller. Of course one does not often get shrinkage and swelling as gross as ten per cent but as little as one or two per cent can be sufficiently troublesome. Paint and varnish slow down moisture changes in wood but they do not prevent them for no paint is impermeable to water vapour.

Even indoors, the relative humidity is changing all the time, especially between night and day. Floor boards and furniture tend to follow the humidity and this is the reason for the ghostly noises one hears in the house at night. If wood is physically restrained from shrinking when it wants to do so it will split, because it has almost no tensile strength across the grain. If it is physically restrained from swelling when it wants to swell, very considerable pressures are built up. The Egyptian method for quarrying large blocks of stone, such as Cleopatra's Needle, was to outline the shape by means of a stress concentration, in the form of a groove in the surface of the rock. Deep holes were made along this groove into which dry wooden pegs or posts were driven. These wedges were then supposed to have been soaked in water until the rock split along the required line.

The shrinkage of cordage and textiles is much the same in

principle as that of wood. Individual fibres change their thickness but not their length with moisture changes, and it is the helical geometry of ropes and textile yarns which causes rope and cloth to get shorter when it gets wet. Flax sails, especially, were very porous and sailing ships 'in chase' would wet their sails to swell the fibres and reduce the porosity.

As we see, the most important effect of moisture on wood is to cause it to swell. A rather less important effect, from the practical point of view, is to change the mechanical properties. Thoroughly wet wood has something like a third of the strength and stiffness of completely dry wood. Biological materials always operate in the saturated state: this gets rid of the problem of shrinkage and swelling at the expense of a reduction in strength. In engineering, cellulose is never used in the completely dry condition so that the range of strength and stiffness is not quite as bad as it sounds.

Wet wood is rather easier to bend than dry but the principal agent for bending wood is heat. Traditionally, wood which has to be bent for tennis racquets and boat ribs is steamed. It is often supposed that the steam does something because it is steam. In fact the steam is a convenient way of heating the wood without drying it out and the mechanism is exactly the same as that used by hairdressers for curling hair. Sometimes amateurs wrap wood in hot wet rags when they want to bend it. The wetness of the rags does not accomplish much but the wood gets heated and the rags may insulate the hot wood and prevent it from cooling too quickly. Wood will not come to much harm in moist heat below about 140° C. but, of course, in dry heat it will soon crack due to shrinkage.

Seasoning

A great deal of rubbish has been talked about the seasoning of timber by craftsmen and by romantic but ignorant amateurs. Wood, as we have said, consists of closed tubes which, in the living tree are partly full of water, or rather sap. In freshly felled wood the moisture content varies but may be over 100 per cent of the weight of the dry wood substance. About 25 per cent of this water is absorbed in the hydroxyls of the fibre wall, the remainder is liquid water inside the cell. Seasoning consists in removing most

of the water in a controlled way: essentially it is a drying operation and nothing more. It is necessary to bring the wood to a
moisture content which is nearly at equilibrium with the environment in which the timber is going to be used for if this is not done
one must expect warping and shrinkage. For external use a
moisture content of perhaps 20 per cent may be suitable, for an
unheated building about 15 per cent, and for a steam-heated
environment about 8 or 10 per cent.

Since the cells are closed, spindle-shaped tubes, the liquid water
inside them is not very easy to get out. It can only be dried out by
diffusing it slowly through the tube walls. This would present no
great difficulty if one were dealing with a single cell but real
lumber contains many thousands and it is necessary to diffuse
the water from the inner cells through the walls of most of the
other cells which lie between them and the outer world. To do this
it is necessary to maintain a moisture gradient between the inside
and the outside of the wood. The sharper this gradient is, the
faster moisture will be lost from the inside. On the other hand if
the moisture gradient is too steep the outside will be notably
drier, in the intermediate stages of seasoning, than the inside and
so it will shrink more and will thus split. This is why one cannot
season too fast without ruining the timber. Traditionally, wood
was seasoned 'naturally' in the open air or in open, unheated
sheds. This might take a year or so for planks an inch or two thick
and seven years for large oak ship's timbers. With primitive
methods and knowledge this is about the best that one can do. One
of the reasons why the better shipyards and coachbuilders were
expensive was that they kept large stocks of valuable timber
seasoned and seasoning.

A great deal of technological work has been done recently on
the seasoning of wood, and safe accelerated drying schedules
have been worked out for all kinds and dimensions of timber. By
carefully controlling the drying rates in large kilns the time for
seasoning can be reduced to a matter of days or weeks. Another
factor which reduces seasoning time is the modern tendency,
because of the existence of efficient glues, to use timber in much
smaller sizes, which of course dry more quickly. Timber which
has been properly kiln-seasoned (which needs expensive kilns and

Plate 1

Chapter 1

Wells Cathedral. It has been necessary to insert these elaborate arches so as to prevent the structure from collapsing inwards.

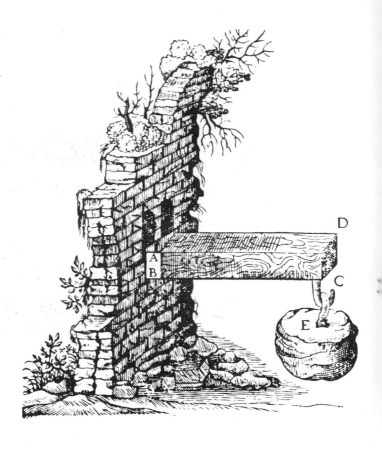

Plate 2

Chapter 1

Cantilever beam loaded at one end. Woodcut from Galileo.
Discorsi e dimostrazioni matematiche (Leyden 1638).

S.S. Schenectady, designed, more or less, on simple beam theory. A crack has started at the sharp corner of a hatchway on deck and has run down to the keel.

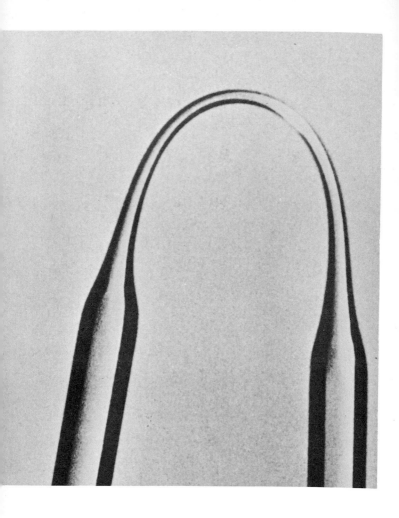

Plate 4

Chapter 3

Glass and other solids when truly free from cracks and defects can exhibit enormous strengths. This silica rod is bent elastically to a strain of $7\frac{1}{2}$ per cent, i.e. a stress of 5000 MN/m². (The normal strength of glass is about 100–200 MN/m².)

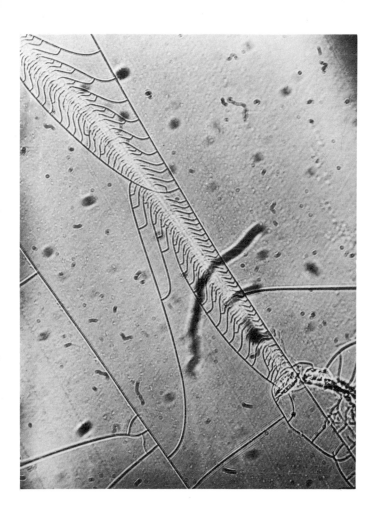

Plate 5

Chapter 4

The strength of brittle solids such as glass is dramatically reduced by surface damage. Even slight contact can cause serious abrasion. This is a photograph of cracks caused by slight accidental contact on the surface of Pyrex glass. Magnification $700\times$.

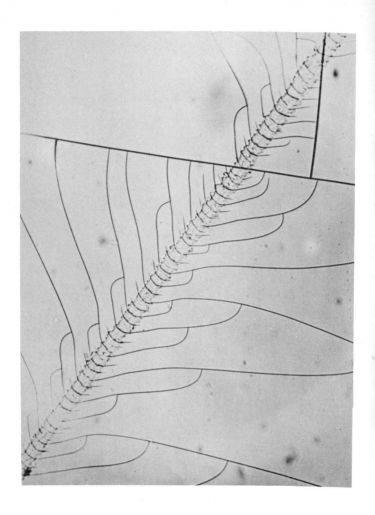

Plate 6

Chapter 4

Cracks resulting from deliberate scrape by a needle-point on the surface of a microscope cover-glass. Magnification $1000\times$.

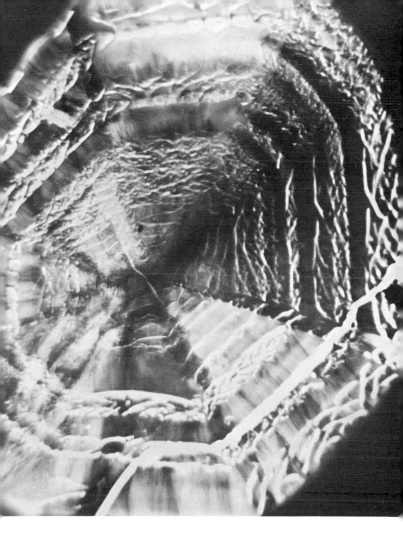

Plate 7

Chapter 4

The strength of glass can also be much reduced by 'de-vitrification', that is to say by the formation of small local crystals. The material contracts during the process of crystallization and frequently cracks in doing so. These cracks often spread from the crystalline into the glassy area. This is a photograph of de-vitrification in silica glass. The central area of the picture is crystalline and much cracked. The black areas in the corners of the picture are glassy material into which the cracks are penetrating.

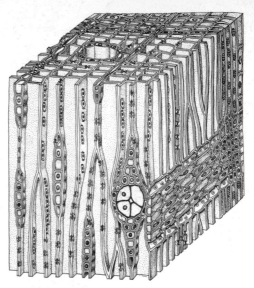

Plate 8

Chapter 6

The cellular structure of wood greatly enlarged. The grain runs vertically. The darker area further from the camera represents 'summer wood', that near to the camera 'spring wood'. A medullary ray can be seen on the picture near the bottom.
Magnification $100 \times$. Courtesy of E.S.T. Mondadori.

Plate 9

Chapter 4

Tin whiskers growing spontaneously on the tin-plated surface of a radio component. This is a frequent cause of faults in electronic equipment.

Plate 10

Chapter 4

Whiskers and other crystals frequently grow by the addition of thin growth layers each of which terminates in a sharp 'step'. These growth steps on a large whisker crystal are all moving towards the bottom of the picture.

Plate 11

Chapter 5

Effect of weak interfaces in stopping or hindering cracks. Material on the left contains numerous interfaces, that on the right none. This crack-stopping mechanism is important in artificial composite materials such as fibre-glass and in wood and other biological tissues. These photographs of a copper-tungsten model material were taken by G. Cooper of Cambridge University.

Plate 12

Chapter 6

Whiskers or needle crystals of hydroquinone growing from solution in water. Note the blurred image of a whisker which has broken away from its mechanical restraints and is in the act of straightening itself. Magnification 1000×. (The parallel stripes are diffraction fringes and are my fault!)

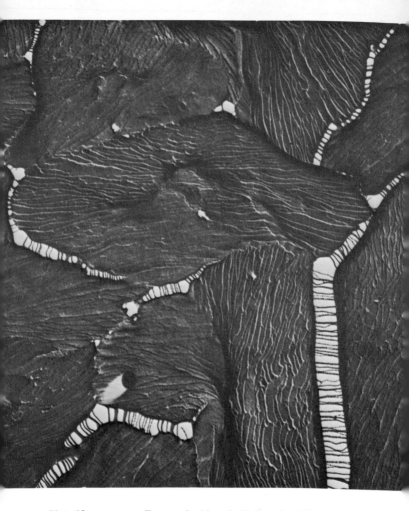

Plate 13

Chapter 6

Fracture in thin polyethylene sheet. The grey plates are crystals of folded polyethylene chains. The new cracks are still bridged by strings of polyethylene molecules which have been pulled out of the crystals. Hence the toughness of polyethylene. Electron micrograph 12,000×. Courtesy of Professor Sir Charles Frank.

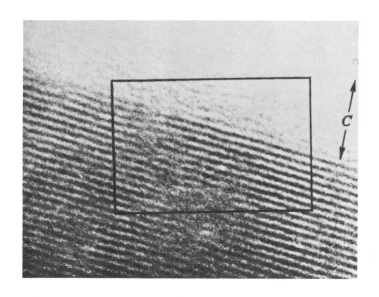

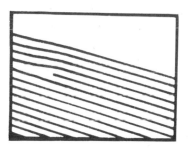

Plate 14

Chapter 9

The dislocation is the most important kind of defect in crystals, especially in metals. This is the first direct photograph of an edge dislocation. The large size of the platinum phthalocanine molecule enabled the lattice spacing of the crystal to be resolved in the electron microscope. Magnification 2,000,000 ×. Courtesy of Sir James Menter.

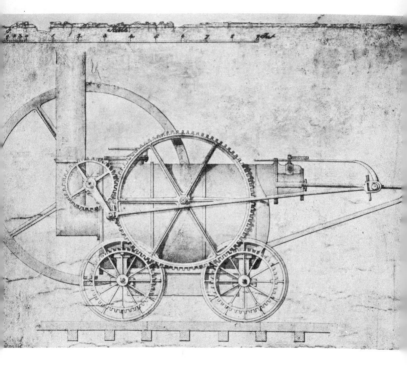

Plate 15

Chapter 10

Trevithick's 1805 high-pressure locomotive. This was a failure because it broke the cast-iron rails. Note that it is shown here upon wooden rails.

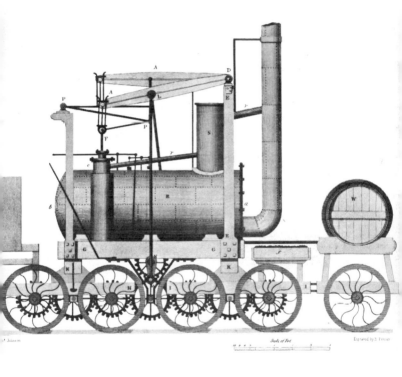

Plate 16

Chapter 10

Attempts to spread the load by using eight coupled wheels. Note that there is no springing except such as is afforded by the flexible spokes. Like other engines this one was converted back to four driven wheels when wrought-iron rails were introduced.

Plate 17

Chapter 10

Rack systems were popular in attempts to solve the
problem of getting sufficient adhesion without putting
much weight upon the cast-iron rails.

close supervision) is in no way worse than 'naturally' seasoned timber and indeed is rather less likely to have picked up the infections of rot during the seasoning process. However, original or commercial sin keeps breaking in and there is undoubtedly a great deal of badly seasoned wood on the market.

The moisture content of wood may be determined simply by weighing a small sample before and after oven drying. In industry it is usually done with portable meters which measure the electrical resistance between two needles pressed into the wood and thus give the answer much more quickly.

Up to about 25 per cent moisture content the whole of the water in wood is held in association with the hydroxyls in the cell walls. At about 25 per cent moisture content however these hydroxyls become saturated and the cell walls can absorb no more water; this is known as the 'fibre saturation point'. Up to the fibre saturation point the lumen or hollow part of the cell is empty of water, above the fibre saturation point virtually the whole of the additional moisture exists as loose liquid water within the lumen. All the dimensional and mechanical changes in wood which are due to moisture occur below the fibre saturation point, that is between 0 per cent and 25 per cent moisture content. After that no further swelling takes place and the additional water simply adds, very considerably, to the weight of the wood.

Wood substance has a specific gravity around 1·4 but freshly felled timber floats (unless it is a very dense species) because, even in the unseasoned wood, there is a good deal of air. Wood will, however, eventually become waterlogged and sink though, like seasoning, this takes a considerable time. The crew of the Kontiki raft were worried lest their balsa logs should sink under them on their long voyage although, in the event, the soakage was not very great. The American clipper ships of the 1850s, the famous 'soft-wood three skysail-yarders', became water-soaked within about ten years by which time they had no doubt paid for themselves very handsomely. The hardwoods from which English ships were usually built are more resistant to soakage and there are several instances of wooden ships afloat and in service for over a hundred years.

Rot

Rot is caused by fungi which live parasitically on cellulose since fungi have no chlorophyl and cannot photosynthesize sugars for themselves. The spores of various fungi are nearly always present in woodwork, just as the germs of many diseases are present in our bodies, but they do not become active unless the conditions are favourable. Rots cannot flourish if the moisture content of the wood is below 18 per cent although the spores can remain alive in quite dry timber, waiting for a rainy day. Even when the moisture content rises above 18 per cent the fungi may not grow if the ventilation is good. As the moisture content of the wood in an unheated structure may be around 15 per cent, it only requires a small amount of damp in an unventilated corner to get the rot going. One cannot always control moisture content but one can generally arrange for ventilation and this is usually a sufficient preventative.

Many chemical treatments are effective in killing active fungi in wood but, in an old and complicated structure, the practical difficulty may be to reach the diseased parts without pulling the whole structure expensively to bits. If the rot is accessible one can generally arrange for ventilation anyway.

In the cyclic natural scheme of things some kind of decay is essential, otherwise not only would the earth be cluttered with the stems of most of the plants which have ever lived, but most of the world's supply of carbon would be locked up in cellulose so that life could not be carried on. This is a general objection to the use of biological materials by man for nature's planned obsolescence may be in conflict with ours.

Wooden ships

'*But not long after there arose against it a tempestuous wind called Euroclydon. And when the ship was caught, and could not bear up into the wind, we let her drive. And running under a certain island which is called Clauda, we had much work to come by the boat: which when they had taken up, they used helps, undergirding the*

*ship; and, fearing lest they should fall into the quicksands, struck
sail, and so were driven.'*
  Acts of the Apostles, Chapter 27.

The wooden sailing ship was *par excellence* the artifact which
made the expansion of Western civilization possible and thus,
more than any other device, was responsible for our present
condition. Wooden sailing ships explored the world and later
surveyed it. They carried passengers and troops, emigrants, con-
victs and slaves. They carried gold and coal, machinery and
books, tea and wool, cotton goods and cheap tin trays, not only
to the ends of the earth but also round the coasts and up the rivers.
For hundreds of years the ship of the line was the ultimate
argument of kings, frequently used. Ships like this are not things
of a dim past, there were first-class passenger sailing ships on the
Australian run within living memory* and there are Admirals
alive who first went to sea in wooden sailing ships.

Although about the middle of the nineteenth century large
improvements were made in both the hulls and the rigs of ships,
for three or four hundred years before that the basic methods of
construction remained nearly constant. The two controlling facts
were that wood swelled and that metals were expensive.

Large ships were heavily framed from 'grown' timbers. That is
to say the curved members, such as ribs, were built up of naturally
curving wood, chosen to have the right shape. The watertight skin
and deck were put on over this closely spaced framework of ribs
and beams in the form of planks, nearly as thick as they were
wide, which ran longitudinally at right-angles to the ribs. The

* The last major wooden (actually composite) *passenger* sailing ship seems
to have been *Torrens* which ceased carrying passengers to Adelaide in
1903. Iron and steel sailing ships of various lines were carrying first-,
second- and sometimes third-class passengers to Australia well into this
century. Contrary to modern popular opinion these were splendid ships
with excellent accommodation and many modern conveniences for both
crew and passengers. Steamship competition was the cause of many hard-
ships in cargo sailing ships but passenger sailing ships had to compete by
offering more comfort. See for instance *Painted Ports* by A. G. Course
(Hollis and Carter, 1961). The largest wooden, sailing, cargo ship – the six-
masted schooner *Wyoming*, of about 6,000 tons burden – was built in
America in 1910.

planking and the underlying ribs thus formed a rectangular trellis with no diagonal bracing or shear members. The edges of adjacent planks were not fastened together mechanically but stood open so as to form a V-shaped groove.

Into this groove oakum, made by picking old rope to pieces in the prisons and workhouses, was driven by means of a mallet and a caulking iron which is a chisel-like tool with a groove along the edge. Outside the caulking there remained an open groove between the planks nearly half an inch wide. In the case of decks this had to be 'payed' which was done by running in hot pitch from a special ladle. When cold the surplus pitch was sufficiently brittle to be scraped off, leaving those pleasing black lines in the deck.* The bottom and topsides were payed or stopped with a putty-like composition. The point of all these arrangements was that the flexible caulking could accommodate shrinking and swelling of the planking, and to some extent movement of the hull, without leaking very much.

The whole structure was, and to some extent was intended to be, quite flexible, almost like a basket. Besides accommodating the shrinkage and swelling of the skin planks, it was supposed, perhaps correctly, that the flexibility of the hull contributed to its speed and sea-kindliness; certainly the Viking ships and the Polynesian canoes were even more flexible. When the much more rigid 'composite' constructions came in in Victorian times one or two of the racing clippers were built with hulls of deliberately controllable rigidity. Of one such ship her rivals would say, as she drew ahead, 'They've unscrewed the beams and we shan't see her again today.'

This was all very well and most wooden ships were watertight in harbour but, without exception, they all leaked when they got to sea. The rate of leakage varied from 'enough to keep the bilges sweet' to something very serious indeed. In spite of all the centuries which he had to learn about it the traditional shipwright seemed to be unable to understand about shear. Any shell structure subject to bending and torsion puts heavy shears into the skin and bending and torsion are just what a ship, especially a

---

*The Devil, incidentally, was a particular seam and the correct expression is 'the Devil to pay and only half a bucket of pitch'.

sailing ship, receives at sea. The orthodox ship construction was like a five-bar gate without the diagonal member.

Since there was no official way of taking the shear it was taken, unofficially, by the caulking which was squeezed and relaxed alternately, like a bath sponge. Occasionally, but surprisingly rarely, the labouring ship spat the caulking from some underwater seam, in which case she probably foundered. More often she just leaked and leaked and leaked. The danger was then, not that she would sink immediately, but that the crew would become exhausted from continual pumping and general misery, after which anything might happen.

When the situation became intolerable an attempt might be made to undergird or frap the ship by passing cables under the hull as St Paul describes in the Acts of the Apostles.* It has been done repeatedly since and very likely, in some corner of the ocean, an Arab dhow is being undergirded at this moment. The whole point of the undergirding cables was to provide some shear bracing and so unless the operation was done with a knowledge and accuracy which were rather unlikely in the circumstances, so as to get the cables roughly at forty-five degrees, the expedient probably had usually as little effect as it seems to have had upon Paul's ship.

As far as the Royal Navy were concerned the nuisance of excessive leakage was largely put a stop to when Sir Robert Seppings (1764–1840) introduced diagonal iron bracing into wooden hulls about 1830. Seppings, who used to say 'partial strength produces general weakness', seems to have been one of the first Naval Architects to have a clear mental picture of the stress systems in a ship's hull. In the merchant service wooden hulls were to a considerable extent replaced by composite and iron and steel construction after the middle of the century. A number of wooden ships continued to be built without adequate shear bracing, however, and such ships got more leaky as they got older, until, in an age when most of the pumping was done by hand, it became uneconomic to run them any longer. Up till 1914

---

*The word translated as 'undergirding' is ' ὑποζώννυμι in the Greek testament and it is quite clear that the operation was familiar both to the ancients and to the translators of the Authorized Version.

Norwegian shipowners were still making money by buying up British sailing ships and running them with windmill pumps.

In spite of their faults, wooden sailing warships were in use for between three and four hundred years and were abandoned by Admiralties with reluctance because they were, in their context, most effective and economical weapons. They had a range and endurance, an independence of overseas bases and an ability to vanish indefinitely into vast spaces which we have only lately regained with the atomic submarine.

Fleet actions were rare, and strategic pressure was generally exerted by blockade and by the threat of a 'fleet in being'. However, until the middle of the eighteenth century it was considered impracticable to keep fleets continually at sea throughout the winter because of the severe and rapid deterioration in the condition of the ships. However, these difficulties were overcome by the efforts of devoted officers. To anyone familiar with the coasts, with sailing ships and with the cellulose molecule, the maintenance of the blockades of Brest and Toulon, winter and summer and in all weathers, must appear as an almost incredible feat. 'Those far-distant, storm-beaten ships, upon which the Grand Army never looked, stood between it and the dominion of the World.'*

Rope and spars came mostly from the Baltic states and the convoys got through with difficulty. Although the blockading squadrons very rarely saw the French they had daily and hourly to struggle with rope and canvas and timber which stretched and broke and rotted. Nelson wrote 'I have applications from the different line of battle ships for surveys on most of their sails and running rigging which cannot be complied with as there is neither cordage nor sails to replace the unserviceable stores and therefore the evil must be combated in the best manner possible.' In spite of this, Mahan wrote 'For twenty-two months Nelson's fleet never went into port, at the end of that time, when the need arose to pursue an enemy for four thousand miles, it was found massed and in all respects perfectly prepared for so sudden and so distant a call.'

* Captain A. T. Mahan, *Influence of sea power upon the French Revolution and Empire*, 1892.

When a sailing ship has a fair wind, even though it be a gale, the loads in her rigging are moderate. However when she is heeling and lurching her way to windward the aggregate of the tensions in the shrouds and stays which support the masts is comparable to the ship's displacement and may thus amount to several thousand tons. Until the middle of the nineteenth century the whole of this load, equivalent to the weight of many railway trains, had to be carried by hemp ropes which were always shrinking and swelling, rotting and stretching so that it called for great skill to avoid the loss of some or all of the masts and spars. There was therefore an understandable reluctance to undertake regularly long voyages to windward in rough weather. A voyage round Cape Horn, for instance, was quite different in character to the routine voyages to the East coast of the Americas or even to India. Bligh's crew, for example, in the *Bounty* mutinied after one appalling attempt to beat round the Horn in which the ship could barely be held together structurally. Eventually Bligh had to turn round and run in the other direction, right round the earth, into the Pacific. Bligh, though unpopular, was a superb seaman and, if he could not succeed, probably nobody else could.

Wire standing rigging was introduced into the Royal Navy in 1838. Its adoption by the merchant service seems to have been fairly gradual (and was not complete until the 1860s) because about this time hemp rope was improved by being laid up, or twisted, mechanically, and thus much more tightly, so that the creep was considerably reduced. By chance the introduction of better rigging more or less coincided with the gold discoveries in California. About half of the emigrants and all of the heavy cargo went by sea and by then numerous clippers were prepared to beat regularly from New York to San Francisco in a hundred days. In the years 1849 and 1850, 760 sailing ships beat round Cape Horn carrying between them 27,000 passengers. It is difficult to determine what proportion of these ships had wire rigging and which used the improved hemp but in any case it is clear that the West was largely won with better rope.*

Another important development was in the matter of chain

* Incidentally wire mining-ropes were introduced in Germany in 1830 and did much to make deep mining safe and economic.

cable. Hemp anchor cables have certain advantages but their drawback lies in the space needed to stow them; the enormous ventilated cable tiers in H.M.S. Victory are impressive. Chain, which was introduced in 1811, could be stowed in a small damp locker and so it can almost be said that chain cleared the space needed below for engines and coal bunkers.

When ships were slow and there were no dockyards on the other side of the world, the fouling of ships' bottoms by weed and the attack on timber by boring animals was a serious matter. Both problems were solved to a large extent by the introduction of copper sheathing about 1770. This did more than any other eighteenth-century innovation to increase the speed and range of ships and it was so successful that shipowners were most reluctant to use iron hulls which could not be coppered directly on account of electro-chemical action between the iron and the copper in salt water. Some iron hulls were sheathed with wood and then coppered. This was popular for warships but it made for a heavy hull. Many of the best racing clippers were therefore composite built. *Cutty Sark* (launched 1869) is planked with teak and greenheart bolted to wrought-iron frames, with adequate shear bracing. The bottom was sheathed with a brass alloy called Muntz metal. There are people who consider this arrangement as being the most perfect construction yet devised for ships of medium size. This may well be true but it is unfortunately also a very expensive one.

The fall in the cost of iron and steel plates in the 1870s made composite shipbuilding uneconomic and, towards the end of the nineteenth century most of the world's deep sea cargo was carried by big sailing ships of almost standardized construction with steel hulls, steel decks, steel spars and steel rigging. Such ships were completely watertight and could be manned by small crews. The loss of speed from their rougher bottoms was compensated by the fact that they could be sailed harder than wooden ships in blowing weather. Over the centuries officers had had to nurse their wooden ships for structural reasons and this had provided a certain measure of protection against excessive sail-carrying. Hard driving clipper captains regarded their new ships as unbreakable and, when expostulated with, would reply 'Hell, she's iron isn't she?' Quite a number of iron and steel ships were driven under and lost by this attitude of mind.

Steamships were in a minority until about 1890 and in any case tended to take the shorter voyages. Of course plenty of wooden steamships were built but the tendency was to turn to iron and steel earlier than in the case of sailing ships. This may have been partly because iron hulls resisted the vibration of the early engines better than wooden ones and also because fouling was less of a problem with continuous speeds and shorter voyages. It is the becalmed sailing ship which fouls quickly.

Orthodox wooden construction is still being used today for fishing vessels, minesweepers and yachts of up to four or five hundred tons. For racing yachts it generally provides the lightest of all hulls and it is also the cheapest way of getting a 'one-off' design built. In its cheaper forms it still suffers from the classical trouble of intolerable leakage in bad weather, especially because of the much higher loads put into the hull by modern rigs. It is true that it is possible to get over this by good workmanship and sophisticated construction but then the cost is higher than that of building in steel or plastics.

Chapter 7 Glue and plywood

or mice in the gliders

> '*When all else fails*
> *Use bloody great nails.*'

Until recently the fact that the strength of engineering materials is usually only between one and five per cent of the strength of their chemical bonds was of little practical significance because the joints between the various component parts of structures were so inefficient that even the strength which the material had was scarcely used. Properly made knots and splices in rope are from forty to eighty per cent efficient; that is, the joint has that fraction of the strength of the continuous rope. Nailed, bolted, screwed, pegged or dovetailed joints in wood are far less efficient than this, at any rate as the joints are generally made. In pure strength, apart from their flexibility, the lashings, sewings and bindings used by primitive peoples, and by seamen down to recent times, are more efficient than metal fastenings, indeed sledges are still made in this way. About 1920, flying boat hulls were made by sewing them together with copper wire.

Wood-screws, beloved of amateur carpenters and boatbuilders are the least efficient of all joints. Between the wars the Germans did a good deal of research on nailed joints and also invented new and clever forms of mechanical connectors. This information is sometimes used today in building wooden houses but, on the whole, mechanical connexions of all kinds have been driven out by modern glues which have made the efficient use of wood possible but have introduced problems of their own.

Glue

Various pundits and committees have made a great mystery about gluing. In fact the elementary theory is simple enough, it is the practice of gluing which is difficult. As we said in Chapter 3, all surfaces have an energy, just because they are surfaces, and this applies both to solids and to liquids. If we consider a solid and a liquid separately, each being in contact with air, then of course each surface has its own surface energy. If now the liquid is brought into contact with the solid, so as to wet it, then the energy of the interface between the solid and the liquid will be less than the sum of surface energies of the two surfaces when they are in contact with air. Thus wetting involves a reduction of energy and will take place whenever the liquid is given the opportunity to meet the solid.

After it has wetted the solid surface, the liquid may be hardened, by freezing it or otherwise. If this is done, the energy of the interface will not be greatly changed and to remove the hardened liquid mechanically will now need strain energy and therefore a mechanical force. Thus, in principle, adhesion is very like cohesion and there is no great difference between the stick of a glue and the strength of a solid. The surface energies of glue-solid interfaces are generally rather less than the free surface energies of strong solids, but not very greatly so. In fact this difference is not usually of much consequence because in both cases the practical strengths are well below what they ought to be. The causes of weakness in adhesion are rather less understood at present than they are in cohesion but no doubt they are generally similar in character.

Any two solids can therefore be glued together if we can find a liquid which will wet them both and then harden. The difficulties which arise are practical ones. Wood can be glued very well by wetting it with water which is subsequently frozen. Such joints are said to pass most of the tests in the specifications for wood adhesives. Hide glue or carpenter's glue may be considered as a variant on ice in that the melting point is raised to a more practical temperature. Hide glue is the same as table jelly, except that less water is added to the gelatine, which may be made from hide, hooves, bones, fish and so on. A stiff solution of gelatine melts in a heated glue-pot at 70° or 80° C. After it is painted onto the wood it sets very quickly as it cools and the joint is soon firm. Unfortunately the process is easily reversed by heating the joint or by soaking it in water. Gelatine is also excellent food for fungi and bacteria. For these reasons hide glue or Scotch glue is only suitable for use indoors. Nevertheless it was used in the early aeroplanes and an attempt was made to protect it by binding the joints with varnished tape which was never very effectual. Until lately, fishing rods were glued with hide glue which was hardened and rendered partially waterproof by soaking the rods in formalin. This treatment was of no use for aeroplanes because it could only be used when the component was small enough for the formalin to penetrate.

Bad as they were, gelatine glues were better than the alternative, the gums or starch glues which were made by boiling some kind of flour in water. Rather strangely, a much better glue had been known but unregarded for centuries. Casein was used as a glue in ancient Egypt and as a vehicle for pigments by the medieval painters. In modern times it was used as a glue from about 1800 in Germany and Switzerland. Why it was not introduced into engineering earlier is not clear. Casein became a recognized adhesive in aircraft and yacht work some time around 1930. Its use made the modern wooden aeroplane, and the Bermuda rig in yachts, practicable.

Casein is the whey in milk and is therefore similar to cheese. Whey is soluble in alkaline water but not in acid. It is thus precipitated from milk by any weak acid, in the nursery notably by rhubarb juice, in industry usually by weak hydrochloric acid. The precipitated whey can then be redissolved in water containing, say, a little caustic soda. On the other hand, casein will react slowly with lime to form an insoluble chemical, calcium caseinate.

Casein glue, therefore, is sold as a white dry powder consisting of dried whey, caustic soda or something similar, and lime. When the powder is mixed with cold water it first of all dissolves to a creamy white paste and then slowly sets to a hard solid. As a glue it is easy to use and nearly foolproof. The only thing that upsets it is leaving the lid off the tin for long periods during storage when the reaction to calcium caseinate takes place prematurely because moisture gets in.

Casein joints set hard in a day or two and are more or less waterproof. However, although calcium caseinate is not soluble in water it does soften to some extent after soaking. Casein glues were very widely used during the war in aircraft and one day somebody discovered that the tensile strength of specimens of wet casein is one-fifth of that of dry casein. It was not un-reasonably feared therefore that the strength of wet aeroplanes glued with casein might also be one-fifth of the strength of dry aeroplanes. Because of the resulting flap we got hold of a hundred similar wooden casein-glued tailplanes. Fifty we sunk in a pond for six weeks, the other fifty we kept dry. In the meantime a jig had been prepared such that the tailplanes could be broken under loads not dissimilar to those they would meet in the air. When all was ready we broke the lot. To our surprise and to many people's relief, they all broke at about the same load, the strength of the wet and the dry ones being very much the same.

The reason for this happy result is instructive. The stress distribution in a glued joint is very far from uniform and, in a typical joint, such as Figure 1, virtually the whole of the load is

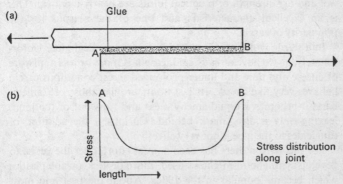

Figure 1. The stress in a glued joint is very unevenly distributed and nearly all the load is transferred through the ends of the joint.
(a) Glued lap joint.
(b) Distribution of stress along joint.
A and B show points of maximum stress.

carried in the extreme ends or edges of the joint. Very little load gets across through the middle. This is merely a variant of our friend the stress concentration which governs so much of engineering and materials science. As a consequence, incidentally, the strength of a glued joint does not depend upon the area of the joint, but mainly upon its width. This is just as true of a mechanical joint as it is of a glued joint and most of the load in such joints comes upon the first and last bolts or rivets. This is one of the reasons why a designer's life is a hard one.

Now dry casein is a hard, brittle substance which transmits loads in the best style of Mr Hooke. When the stress at the ends or edges of the joint reaches the strength of dry casein therefore, cracks appear at the edges of the joint which immediately pro- duce their own private local concentrations of stress, and so the cracks run through the middle of the joint, much as they would in glass. Damp casein is very like damp cheese and its elastic behaviour is anything but Hookean. As a result, it just yields in a soggy sort of way to the concentrations of stress at the edges of the joint and the load is shared by the glue in the middle. With a glue like casein the various effects cancel out so that the wet and dry strength of practical joints are nearly identical. This is an excellent characteristic and one of the reasons for the popularity of casein.

In a sterile world casein would be nearly an ideal glue. Unfor- tunately, as we have already said, casein is more or less a mixture of cheese and lime and under prolonged moist conditions casein behaves very like cheese. Its last hours are like those of Camem- bert; it becomes a liquid smelly mess and runs out of the joint, leaving only a dirty mark behind. Curiously, the addition of fungicides to the glue is not very effective.

For this reason there has been a great effort, over the years, to develop synthetic resin glues based, more or less, on the plastics which became popular in the 1930s. About the best and most durable of these, in the early days, was phenol-formaldehyde or 'Bakelite'. This was available in the forms of either a treacly liquid or a dry powder. Under heat and pressure, the powder would melt and both it and the liquid would then set per- manently to a hard, insoluble solid nearly immune to decay. Where hot-setting conditions could be used, phenol-formalde- hyde became the basis of a series of really excellent glues. Since it is essential, not only that the glue should be heated to about 150° C. but also that there should be no appreciable gap or space in the joint, the gluing operation had, in practice, to be carried out in a heated hydraulic press. For this reason it was really only suited to the manufacture of plywood, for which it was immensely successful.

Although phenol-formaldehyde glues resulted in good waterproof plywood, they left the problem of the gluing of joints in the assembly of aircraft and boats unsolved, since it is generally impracticable to heat the joints of large structures in a controlled manner. It is true that phenol-formaldehyde resins can be set in the cold but this requires large additions of a chemical catalyst, or hardening agent. These catalysts were strong acids which damaged both the wood and the workmen.

The first synthetic assembly glues were therefore of the urea-formaldehyde type which could be set with much weaker catalysts. Although successful in their way, the earlier glues of this kind were structurally rather dangerous. When used in a properly fitting, thin joint they were reliable enough but, when the joint fitted badly so that the glue-line was thick, the shrinkages and internal stresses set up when the glue hardened often caused the glue to craze and to fall to bits in the joint. As it was impossible to inspect the inside of the joint after manufacture this was a potential cause of accidents.

Another trouble was that, once the catalyst or hardener had been added to the glue, setting began, so that the safe working life of the liquid glue was a matter of minutes. Moreover there was no infallible way of telling how long ago the hardener had been put in. With the usual frailty of human nature this led to all sorts of mistakes. This defect was ingeniously got over by developing a hardener which could be painted by itself on to one half of the joint while the glue was painted on the other half, nothing began to happen in the glue until the two halves were brought together. As a further guard against human frailty the hardener, and occasionally the glue, were dyed characteristic colours.

This is about as far as the development of glues had got by the end of the war. Excellent and highly waterproof plywood was universal. For assembly glues the arguments were about even between casein and urea-formaldehyde. Casein was nearly foolproof in application, of excellent strength both wet and dry, but rotted catastrophically when it got the chance. Urea was resistant to rot but in these early forms was tricky to apply and rather apt to fall to bits without warning in middle life. Since then, urea glues have been much improved and we have two new synthetics in particular, resorcinol formaldehyde and epoxy resins which, though they are rather expensive, have nearly all the technical virtues, bearing in mind however that epoxy resins frequently cause dermatitis.

Among the dozens of glues now available the arguments for and against are mostly on the grounds of ease of application, durability and cost. With all reputable glues the joint, when properly made, is stronger than the surrounding wood. Typical failures exhibit a thin layer of wood covering the glued surface. The strength of a well made glued joint in good condition is not increased by nailing or screwing the joint in addition to the glue. On the other hand, all glues need to be tightly clamped while they set and the simplest way to do this is generally to assemble the wet joint with nails or screws; having once done this, there is no particular benefit in taking the fastenings out afterwards.* Furthermore, a joint in poor condition may peel, like a banana skin, and the presence of mechanical fastenings is a great insurance against this. In the old days, with casein, it used to be said of some of the aircraft in the tropics that they were held together by the assembly brads. In most cases this was slander but I have seen instances where it was not far from the truth. Where glues are concerned, I, personally, would not scorn to wear both a belt and braces.

Laminated wood and plywood

One of the troubles with wood has always been getting it in the right sizes and making sure that it is free from hidden defects. Years ago one could buy great logs of Kauri pine from New Zealand, yellow pine from Quebec and so on which were virtually perfect but those days are long gone. Nowadays most of the wood which is used in engineering is laminated in one way or another. This means that it is cut up into comparatively small pieces which are then glued together again, usually in hydraulic presses with synthetic glues. In this way members of any size can be supplied and practically the whole volume of both large and small trees is made use of. Furthermore any serious defect is seen

* See *Structures*, Penguin Books, 1978, Ch. 13.

and rejected. Such members can quite easily be made in curved shapes and a not uncommon obstruction on the roads of England is a lorry carrying vast timber arches for some architectural project. The shortage of high-grade timbers for aircraft and boat-building would have been a serious matter during the War if all these woods had not been 'upgraded' by lamination with perfectly satisfactory results.

These laminated woods were simply ordinary timber cut up and glued together again. There existed, however, a distressing class of material known as 'improved woods' which had very much the properties and fate that one would expect from so hubristic a name. In these materials the wood was impregnated to a greater or less extent with a resin and then compressed to a considerably higher density. The intention was that the mechanical properties would be thereby improved; they were, but generally only in proportion to the increase in density, at the same time much of the toughness of wood was lost. Worst of all, they liked to swell back to their original dimensions in water and this swelling was generally unpredictable and irreversible. To be fair, these materials performed a useful function for a time in the propellers of Spitfires and similar aircraft.

Plywood is a different story and can almost claim to be a new material and a most successful one. It is made by gluing together three or more veneers, or thin sheets of wood, with the grain directions crossed. Veneers are made either by slicing or peeling. In the case of sliced veneers a baulk of timber is clamped and slices are peeled off it by a machine very like a large plane. With peeled veneer, a round log is heated in a steam pit for about twenty-four hours and then set up and rotated in a lathe in which a long knife peels the veneer circumferentially at a speed which is wonderful to watch. The veneer is cut up, dried, the defects are removed and finally stacks of plywood are glued in large presses.

The early plywoods were glued with vegetable or blood glues and, having virtually no moisture resistance, the word became almost a term of abuse. The introduction of phenolic glues changed all this and, incidentally, presents an interesting picture of the way in which the public image of a material can be changed. Modern phenolic glued plywood is quite impermeable to water, in the sense that the veneers will not come apart when it is soaked, and it has become an important material in modern boatbuilding.

As one would expect, the dimensional movement due to moisture changes is about halved. In other words the maximum movement, which now occurs in both directions, is around five per cent and in practice generally a good deal less. However, when the surface veneers dry, as in the hot sun, they are subjected to cross-grain tension and may thus 'check', that is produce a large number of small cracks. These, in themselves, do not do much harm, but in unpainted ply they form little traps for moisture and bacteria and so lead to trouble. Most of the original infections in plywood are killed in the hot pressing process, but if the material is subsequently exposed to bacteria or fungi and water it will rot quite quickly.

Wooden aeroplanes

It is generally a mistake to despise any form of construction and this certainly includes the 'stick and string' biplane. The main factor which governs the choice of materials and structural form is the ratio of the load on the structure to its dimensions. When the loads are comparatively small in relation to the size it is generally best to concentrate the compression loads into a few compact, rod-like members and to diffuse the tensions into fabric and cords. This is clearly the case in the rigs of sailing ships, in tents and in windmills and with certain variations it is true in balloons as well. Any other arrangement would be heavier, more expensive and inconvenient.

All the early aircraft had very low wing loadings for sufficient reasons. Their actual dimensions were, in many cases, not much smaller than those of equivalent modern aircraft, but their weights were less than a tenth of modern, hard-skinned machines. In the circumstances, a construction of fabric over a wire-braced framework of wood or bamboo was logical and efficient, and sometimes, nothing else would have got off the ground with the power available. The biplane form enabled efficient lattice girders, and also an efficient torsion box, to be achieved in a very robust form with little weight. The solid members needed only to take compression and since the principle danger in this condition was buckling, such components required to be as thick as possible for their weight and for this purpose bamboo or spruce was especially suitable. The numerous tension members could be simple piano wires. The problem of tension joints in bamboo, however, was always a serious one.

This philosophy produced strong and excellent aeroplanes so long as one was quite sure which members were in tension and which in compression, for, though a strut could at a pinch take tension, a wire could not take compression. In some of the more elaborate biplanes it was not always easy to tell which way the loads were going and it was a stock joke at one time that the way to check the rigging of a certain aircraft was to put a canary between the wings; if it got out something was wrong. S. F. Cody, of the 'Cathedral' biplane, was addicted to elaborate rigging and furthermore, not very technically minded. My grandfather, who was one of the aircraft pioneers, told me that he had a long argument with Cody about whether a certain member was in tension or in compression in flight. Cody maintained that it was in tension and had provided a wire. It turned out that my grandfather was right for Cody was killed a few minutes afterwards for just this reason. By some kind of irony this was exactly the reverse of the trouble with masonry cathedrals which fell down because they turned out to be in tension when the builders held that they were in compression.

It took some time, and many lives were lost, before the stressing conditions to which an aircraft is subject were sufficiently understood. In this country this achievement was largely due to the group of highly intelligent young men who were gathered together at Farnborough in the famous Chudleigh Mess in the first War. The principles of stressing and testing aircraft have remained much the same from the days of wooden biplanes down to supersonic fighters although there are many differences in practice.

When an aircraft has been designed and built a full-sized specimen must be tested for strength and stiffness. Stiffness testing is relatively simple but strength testing may involve engineering which is both heavy and difficult. In the 1914 period the custom was to turn the aircraft upside down and then to load the wings with bags of sand or lead shot distributed so as to represent the various aerodynamic loads which occur under the worst conditions, such as pulling out of a dive. Quite soon the loads on aircraft got too big for this method (though shot-bags are still used from time to time for certain simple tests) and nowadays the loads are applied by means of hydraulic jacks operating through very elaborate multiple lever or 'family tree' systems; each of the hundreds of branches ends in a mechanical attachment to the wing surface. There are so many of these attachments that the diffuse nature of the aerodynamic load is imitated (Figure 2).

In its better forms, such as the Avro 504 and the various Moths, the wooden biplane was almost everlasting* and the structure

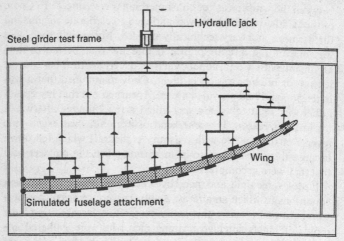

Figure 2.   When an aircraft wing is to be tested, it is mounted by its root fitting in a large steel framework. Many hundreds of attachments, distributed over the wing surface in both dimensions, are used. In this way the effect of aerodynamic pressure is imitated.

*These wooden biplanes are much longer-lived than motor-cars. There are still plenty of Tiger-Moths about – presumably 30 or 40 years old – and only the other day (1975) I saw a de Havilland Rapide – a twin-engined cabin biplane – flying around very happily; it had probably passed its fortieth birthday.

could really only be broken by flying it into the ground. The feeling of structural security about flying in strutted and braced aircraft was very comforting but unfortunately did not always extend to the engines with which they were fitted. Cantilever monoplanes felt much more dangerous.

As loads increased the general trend of design was however undoubtedly towards monocoque, or hard-skinned, monoplanes in which the loads, as far as possible, were taken in the skin. There is no difficulty about taking tension in a thin membrane, the problem is how to take compression without causing the skin to buckle. In practice this was done by a compromise, the thin skin being stiffened by attaching to it spars and stringers, with which it shared the load, the whole rather elaborate surface forming a shell stiff in bending and therefore unlikely to buckle.

The outstanding early example of this was the D.C.3, later known as the Dakota, and this was followed by the Spitfire and by many of the famous aircraft of the last War. All these were metal aircraft, built of aluminium sheet to which were riveted L-shaped aluminium stringers. This system proved almost exactly equal in structural efficiency, that is in weight, to wood and fabric with the advantage of a smoother outside surface and the almost total elimination of airframe maintenance.

This construction proved very successful and, with minor modifications, is still the standard way of making aeroplanes. By 1939 it was widely believed that no more wooden aeroplanes would be built and this might have come true if the War had not created shortages of aluminium and of the machinery and skilled men for handling it. Furthermore there were furniture firms short of work and again the development time for a wooden aeroplane has always been much shorter than that for a metal one.

One expert achieved unwanted fame by stating, categorically, that it was no longer technically possible to build modern aircraft out of wood. The ink was scarcely dry upon this document when the Mosquito appeared. This wooden aeroplane was one of the most successful aircraft of all time, and 7,781 of them were built. It was probably more detested by the Germans than any other Allied plane.

Besides the Mosquito and a large number of trainers, the other big production of wooden aircraft was of gliders. Most of these were large machines with spans up to 110 feet (33 metres), frequently made to carry tanks and other heavy equipment. The original idea was that the gliders should be built to make one flight only. This turned out to be impracticable, partly because they were needed for training and also had to be moved between aerodromes because of changes in the strategic and tactical situation, but also, more importantly, just how do you build an aeroplane for one flight only? In practice the gliders became much like any other aeroplanes, except that they had no engines.

On the whole the wooden aircraft were extraordinarily successful and I suppose that we could hardly have won the War without them. Far too many of them, however, soon started to produce technical problems of one kind or another which almost immediately swamped the tiny number of experts in organic materials who were available at the Royal Aircraft Establishment. It was most fortunate that a young Cambridge biologist called Mark Pryor was extricated by the powers that be from a searchlight unit and sent to Farnborough to take over this work. The fact that the accident rate in wooden aircraft was kept to a reasonable figure and that enough gliders arrived in France in adequate structural condition was due in a considerable measure to Mark Pryor. Quite a number of soldiers and airmen owe their lives to his interminable wartime journeys between his microscope and the aircraft factories and airfields.

Given all the circumstances, it is difficult to see how most of the problems which arose could have been foreseen. The old fabric biplanes were entirely satisfactory but then they were made out of small pieces of wood, they were well ventilated and they were kept in good dry hangars. None of these conditions applied to wartime aircraft. First of all the new machines were of monocoque construction with comparatively heavy spar-booms and stringers rigidly glued to thick plywood webs and skin. We shall come to the further consequences of this construction shortly. Its immediate result was to divide the aeroplane up into a large number of badly ventilated and inaccessible compartments. As the aircraft were left out 'dispersed' most of the time in English or in tropical rain these compartments soon became little damp boxes, often with a puddle at the bottom. Rot, either in the glue or the wood or both, was only too likely within a few months. It is not very easy to arrange ventilation schemes when the designer has forgotten to do so and often the best that could be done was to arrange to leave the inspection doors open when on the ground.

Many aeroplanes, however, collected loose water inside, often in the most inaccessible places. The cure was to provide drain-holes, not just anywhere, but at the lowest point in each compartment. This was done with almost no result. It turned out that mechanically drilled small holes in plywood were fringed on the inside with a little coronet of splinters which could not be removed because it could usually neither be seen nor reached. These splinters soon collected enough dirt and fluff to block the drain holes and there we were with a puddle again. The cure for this turned out to be to burn the drain-holes with a red-hot skewer which, of course, gives a clean edge to the hole. Obvious, when you have thought of it. This solution was applied both to aircraft and to Motor Torpedo Boats, which had much the same troubles as wooden aircraft. The burnt drainage holes did a lot of good, but there seemed to be no cure for mud. Mud is apt to be thrown up by aircraft wheels in landing and taking off, muddy water enters the aircraft through drainage holes and all sorts of other orifices and then the water drains out, leaving a film of wet mud. This often contains the seeds of grass and other plants which before long sprout, like mustard and cress on a damp flannel. Internal gardens do no good, of course, to aeroplanes.

On the whole these troubles were less severe with powered aircraft than they were with gliders. Powered aircraft naturally flew more frequently and the consequent draughts were good for the aeroplane and bad for the fungi. More and more gliders were being produced all the time and it was quite impracticable to house them all in hangars, so they stood endlessly around the edges of airfields in the rain, waiting for an invasion which might come this year or might come next. As rather over 5,000 gliders were built in this country it was not possible for 'experts' to inspect them all continually so Mark Pryor issued instructions that he should be informed if they stank.

Now an unpleasant smell in a wooden structure is due to one of three things; drains, mice or rot and they all smell very similar. Drainage smells arose because there weren't any drains and could cause damage. Mice entered aircraft in pursuit of sandwich crumbs, usually under the floorboards. By the time they had eaten up the crumbs they had forgotten the way out and in their hunger ate the insulation from the wiring. Mark dealt with mice by obtaining an official issue of cats. Rot was more complicated and difficult. In all the circumstances some rot of some kind was almost inevitable in a good proportion of gliders. Perfectionism is out of place in war and the practical problem was to condemn and ground those gliders which were dangerously attacked and to detect and stop the rot in those in which the attack was trivial. This called for a great deal of judgement because there are about forty different species of rot, and the damage which they do varies and is not necessarily proportional to the visible effects.

Problems of rot were always with us but there were other problems as well which were just as serious. As I have said, the general structure of these aeroplanes was quite different from the old fabric biplanes. The main spar booms and other main structural members were sizable pieces of laminated wood, several inches square, and were generally boxed in on three sides by the plywood skin and shear webs. Now the spruce spar boom wanted to shrink and swell about twice as far as the plywood which was glued to it and this naturally gave rise to serious stresses near to where the two met along the glued joints (Figure 3).

Large pieces of timber take some considerable time to come to equilibrium with the surrounding humidity and, because the English weather changes so often, there was generally no time to build up dangerous differences in swelling strains so that we had comparatively little trouble from this cause, so long as the aircraft were in this country. When they were sent overseas the situation was different. In many climates there are long dry seasons followed by long wet seasons, each season giving ample chance for the wood to dry out thoroughly and then, in due time to soak up a great deal of water and swell. In such places there

(a)

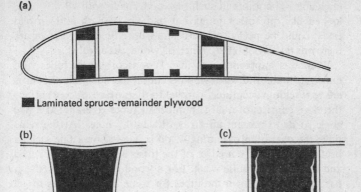

■ Laminated spruce-remainder plywood

(b)   (c)

Figure 3. Effect of moisture changes upon the structure of a wooden aircraft wing.
    (a) Diagrammatic cross-section of typical World War II wooden
       wing.
    (b) What happens when the spruce spar booms shrink more than
       the plywood skins and webs.
    (c) The final result – fracture of spruce spar boom or else glue
       failure.

was serious trouble. Big stresses were built up near the glue-lines;
if the glue was in bad condition it broke; if not, the wood failed
near the glue. There was really no cure for this except to bring the
aircraft home.

Gluing troubles arose, not only from the glues themselves but
from other causes. The worst of these was the so-called 'case-
hardened' failure. It will be realized that there is no way of
testing a glued joint which actually forms part of an aeroplane
except by breaking the aeroplane, which is a self-defeating
activity. One is therefore guided to a considerable extent by the
appearance of the joint and also relies upon the inspection procedure during manufacture.

Soon after wooden aircraft went into large scale production it
began to be realized that a proportion of aircraft plywood was

ungluable. The joints in such plywood, made with all due care, looked like any other joints but had no strength and, in bad cases, could be peeled off with one's little finger. Worst of all, there was no way of telling which plywood was affected.

What was happening was this. Wood consists of tubes with quite thin walls and when wood is cut on the side grain the tubes are very seldom accurately parallel to the surface. Such a surface therefore consists of a large number of tubes emerging at a fine angle so as to present an array of slanting holes. At the same time, the operation of cutting wood is, viewed on a fine scale, a brutal one and the cut edges of the tubes are therefore damaged and mechanically quite weak. For a wood glue to be effective it has to penetrate down the tubes for some distance so as to get hold of the undamaged wood. If it is prevented from penetrating, the glue will adhere only to the damaged edges of the tubes which will break away as soon as a load comes on. In the case of 'case-hardened' plywood the edges of the tubes are bent over and turned inwards by the hot plattens of the press which makes the plywood, so that the glue is prevented from penetrating and the joint has no strength (Figure 4).

This is a very lethal condition and has been responsible for a great many accidents and much loss of life. The only reliable cure for it is to sandpaper off the damaged surface of the plywood. The sanding must be done thoroughly, merely scratching the wood is no good. Since one never knows what plywood is affected it is absolutely necessary to sand *all* plywood which may find its way into an aeroplane. This turned out to be a major administrative problem. It proved impossible to rely on hand sanding and a system of mechanical shot-blasting was instituted, after which the plywood was dyed or stamped to indicate that it was safe to use.

Wood is not a material which suffers fools gladly and a great deal of the trouble with wooden aeroplanes was due to wooden people. This could occur at all levels. Some designers felt that wood 'ought' to behave like a metal. If they made a mistake on this account then it was the wood's fault, not theirs. Engineer Officers, or at least the more recently joined ones, were brought up to consider metal the thing and sometimes had no patience

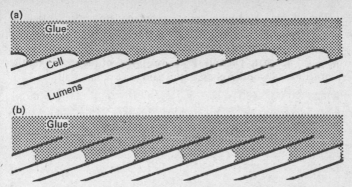

Figure 4. Many accidents could be traced to the use of so-called 'case-hardened' plywood in aircraft. Plywood can be rendered 'un-gluable' as a result of heat and pressure during the manufacturing process. The cure is to sand the surface thoroughly before using the plywood.

    (a) 'Case-hardened' plywood. Emergent edges of cell walls are burred over during hot-pressing preventing adequate entry of glue into the tubes of the wood.

    (b) Properly sanded plywood. Glue is enabled to enter cell lumens in considerable depth and thus to provide a reliable joint.

with wood. There was the Engineer Officer, in civil life a garage proprietor, who lined up his wooden aeroplanes on the tarmac every morning and had them well hosed down.

In the factories, experienced inspectors were few and over-worked. Some mistakes were due to genuine misunderstanding or fatigue; others I am afraid can only be accounted for by criminal stupidity or irresponsibility. There are always a few people for whom the most obvious sequences of technical cause and effect have no meaning. Gluing is not so much a skilled job as a respon-sible one and a large number of mistakes are available to a determined man, all of which can have dangerous results.

Herein, I think, lies the real difficulty about wooden aeroplanes. If they are wanted at all they are probably wanted in large numbers and to be made in a hurry by unskilled labour. Wood, which is really a craftsman's material, does not take kindly to the inevitable abuses of an emergency.

For all these reasons wooden aeroplanes are under a cloud at the moment. However it would take a brave man to prophecy that they will never come back. One never can tell where wood will turn up next. There is now a very efficient motor car on the market with a wooden chassis. It is said to be selling well.

Chapter 8   Composite materials

or how to make bricks with straw

> '*Then the officers of the children of Israel came and cried unto Pharaoh, saying, "Wherefore dealest thou thus with thy servants? There is no straw given unto thy servants, and they say to us 'Make brick': and, behold, thy servants are beaten; but the fault is in thine own people."*
>
> '*But he said "Ye are idle, ye are idle: therefore ye say 'Let us go and do sacrifice to the Lord'. Go therefore now, and work; for there shall be no straw given to you, yet shall ye deliver the tale of bricks." '*
>
> *Exodus*, Chapter 5.

Ever since Pharaoh had labour troubles about putting straw into bricks there have been reinforced materials of one kind or another although they have only come into prominence as strong materials quite recently. It seems probable that the purpose of putting chopped straw into the Egyptian bricks was just the same as that of the Inca and Maya in putting plant fibres into their pottery, that is to prevent cracking when the wet clay was dried rapidly in the sun. Egyptian bricks were not fired in an oven but since it hardly ever rains in Egypt this did not matter very much. Clay, of course, makes a nice plastic paste when it is wet but the shrinkage on drying is considerable and the problem of drying clay is very much like the seasoning of timber. If it is not done slowly the clay will crack. The Egyptian sun is an excellent agent for drying clay but also a very rapid one and a little straw is effective in controlling and reducing the consequent cracks. It is most likely that the reinforcing effect of the fibres on the clay after it had hardened was incidental.

However, even quite small additions of fibre do have a considerable effect in improving the strength and toughness of weak, brittle materials and there are a great many instances of this. For instance it used to be the habit of English builders to add a little hair to household plaster. I remember being told, as a child, by a workman that bull's hair was much better than cow's for the purpose because, of course, bulls were stronger than cows. I have never made any actual experiments with either bull's or cow's hair and so I cannot pass any opinion on the subject but I have put paper pulp into plaster of Paris. The results, which are excellent, are shown in Figure 1. The trend of improvement in

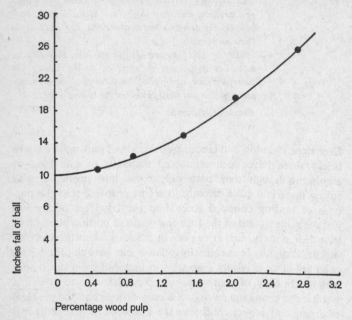

Percentage wood pulp

Figure 1. Brittle solids are often made much stronger and tougher by the addition of quite small amounts of fibre.
Effect of added paper fibres on the impact strength of plaster of Paris.

impact strength, that is to say in work of fracture, is steeply upwards and even very small additions of fibre make a big improvement. Unfortunately, adding fibres to wet plaster very rapidly thickens it and somewhere between two and three per cent of fibre results in a paste which is too stiff to mix. To some extent this difficulty can be got over by using a different type of cement and by consolidating it in a hydraulic press. Figure 2 shows the effect of asbestos fibres in a phosphate cement very like the cements dentists use in one's teeth; the results are much the same except that we can get more fibre in and thus more strength and toughness.

For a time during the Battle of the Atlantic the worst shipping losses were taking place in the central parts of the Atlantic where

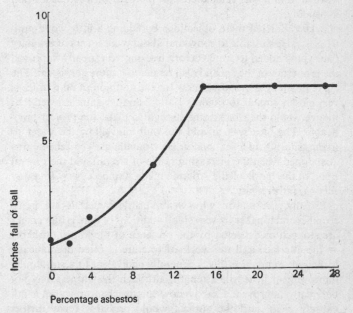

Figure 2. Asbestos fibres have an excellent effect upon the impact strength of a ceramic cement, in this case a phosphate compound. However, no further improvement occurs with fibre additions of more than 15%. I do not know why.

it was not then possible to provide air cover. It occurred to that very eccentric genius, the late Geoffrey Pyke, that one solution would be to tow an iceberg into the middle of the Atlantic and use it as an aircraft carrier. In some ways this was probably an excellent idea, but it was found on investigation that the mechanical properties of natural ice were unsuitable. Ordinary ice is quite weak in tension, it is brittle and cracks spread easily, which is why icebreakers can earn a living, and worst of all, the strength of ice was very variable. An iceberg would probably break up if it were bombed or torpedoed, but, even if it were not attacked, an iceberg large enough to operate aircraft would span at least a pair of Atlantic waves, which run about five or six to the mile in bad weather, and it was calculated that it would then fail as a beam in bending.

Pyke countered these objections by adding a little wood-pulp to his ice. He was able to show that about two per cent of ordinary paper pulp added to water before freezing very greatly improved the properties of the ice and also rendered it more consistent. The curve of strength and toughness for the addition of pulp to ice is very closely similar to Figures 1 and 2 and the sums showed that the ice would then be strong enough to make the project practicable. The idea was to add the pulp and allow the water to freeze naturally in a sea-loch in Newfoundland. The scheme was abandoned when the increasing range of aircraft and the general trend of the Battle of the Atlantic made it unnecessary. In a way this was rather a pity.

Broadly, the reason why weak, brittle materials are made stronger and tougher by very small additions of fibre is that cracks are stopped or deflected by the presence of the fibres, probably at the interface and the work of fracture is often dramatically increased; in practice this is generally equivalent to a substantial improvement in tensile strength, although the stiffness may not be greatly improved. Exactly how this mechanism works is not entirely clear and the subject would probably repay further research. In the form which we have discussed so far, that is to say quite small additions of random fibres to a brittle matrix, there is not much market for it at the moment, perhaps because nobody seems to want sun-dried bricks or mobile icebergs. However I would not be at all surprised if this approach came into fashion again in some different guise. At the moment the mode is for rather a different way of using fibres.

Those who work in the field of fibrous materials receive, in one way and another, a liberal supply of suggestions about schemes which we ought to try. Nearly all these schemes ignore the fact that, in order to compete with the quite efficient existing materials, it is necessary to get a large number of fibres into a given volume of the new substance and this is just where one begins to get into practical difficulties.

The simple systems which we have described contain about two per cent of short fibres, added to a matrix which, at one stage, is more or less liquid. Up to this point the fibres can be incorporated, if necessary, by stirring them in with a spoon. Beyond about two per cent however this becomes impracticable and the operation shows serious signs of getting out of hand. Suspensions of long thin fibres in liquids behave very like solutions of long thin molecules; both have a thickening effect which it is difficult to credit unless one has actually experienced it. The pulp (that is a suspension of wood fibres in water) for the paper from which this book is made had to be pumped and handled at a concentration of about one half of one per cent in order to be in any way manageable.

Papier-mâché

Beyond about two per cent, therefore, it is impossible to add fibres to a matrix and it becomes necessary to add the matrix to the fibres. This naturally changes the whole approach. Nearly always the fibres are packed tightly together in some form such as paper or cloth which is then soaked in a resin or some other binding agent. This form of the idea, too, turns out to be very old, again Egyptian. Elaborately shaped mummy cases were made from papier-mâché. This material is made by sticking pieces of paper together with paste or gum over a mould. When the glue is dry the shell is removed and painted. In Egypt papyrus was used, in the form of old manuscripts. When steamed apart by archaeologists most of the scraps of papyri turn out to be dull stuff, as one might expect, but the process does yield a small trickle of important Greek literature and is probably our only hope of recovering the works of Sappho.

After classical times papier-mâché fell out of use because papyrus fell out of use, but it was revived, almost unchanged, in the eighteenth century, especially in France, for making furniture using paper. It was used in England, until very recently, for advertising models and, during the last War, a great many fuel tanks and other aircraft parts were made in this way. The methods used, down to about 1945, were in no way different from those employed by the ancient Egyptians except that paper was used instead of papyrus.

Since it might be useful, it is worth mentioning how the process is done. A mould of some kind is needed which can be made from clay or Plasticine or plaster of Paris. This mould is then covered with soap or linseed oil (or possibly silicone car polish) to prevent the papier-mâché from sticking to it. Almost any paper will do but soft papers such as used to be used for sugar bags are most suitable. For glue a sloppy mixture of Scotch (hide) glue and office paste (starch) is best. It should be about as thick as pea soup. When all is ready, tear off pieces of paper about the size of the palm of one's hand and dabble them thoroughly in the glue until they are quite soft. Then press them on to the mould. Go on doing this until you have built up the necessary thickness, which may be more than you expect. When the moulding is thoroughly dry it may look rather like something the cat found (but do not despair, no doubt the Egyptians felt the same); however, it is much improved by sandpapering and painting. Many coats of oil paint are needed since the material depends entirely for protection against the weather on paint. The moisture resistance, of course, is bad, but not as bad as one would think. The mechanical strength is surprisingly good. Do not try to use synthetic glues or the result will be brittle, probably because the adhesion is too good.

If the resistance to moisture and fungi had not been even worse than that of natural timber, probably papier-mâché would have been more widely used because technology is always demanding a way of making light, strong, shell-like structures with elaborate curvature for coachbuilding, boats, armour, baths, furniture, luggage, fuel tanks and so on. For centuries, however, the process was stuck for lack of a strong waterproof adhesive and people were impelled to beat heavy, easily dented shells out of metal with great labour.

Moulding powders

However, in 1906 Dr Baekeland discovered that phenol and formaldehyde could be reacted together to make a resin which, though initially liquid or treacly, could be hardened by heat to an infusible, insoluble solid. Baekeland had a mind well open to commercial possibilities and he had already made a fortune by inventing and exploiting 'Velox' photographic printing paper, but even he does not seem originally to have been too optimistic about the applications of his resin, which was originally sold as a substitute for natural resins in lacquers and varnishes. I have been told that the great Bakelite Company started life in Ed-wardian times as the 'Dammard Laquer Company' and that it marketed three grades of varnish, 'Dammard', 'Dammarder' and 'Dammardest'.

Bakelite resin by itself is, when set, a hard, brittle substance of no great strength, much like natural rosin. It was of some use as an ingredient in lacquers, especially as an insulant in the electrical industry, and it was later to prove an excellent adhesive for ply-wood, but it was quite useless for mechanical purposes in solid lumps in the pure form. The turning point occurred when Baeke-land noticed that the addition of fibres to the resin before it was hardened transformed its strength and toughness.

Most of the early applications were as a moulding powder. In this manifestation the partly set resin is mixed with short cellulose fibres, generally wood-flour. The resulting dry powder can be put into a heated steel mould in which the powder softens and, as the pressure of the hydraulic press is exerted, it flows to fill the inter-stices of the mould and then hardens irreversibly. The first com-mercial Bakelite moulding is supposed to have been made in 1916, the gear-lever knob for the Rolls Royce car.

Because of its light weight, low cost and the enormous saving of labour in making complicated shapes, Bakelite mouldings became extremely numerous between the wars and at one time there was even a major project to flood the undertaking market with Bakelite coffins. A well made Bakelite moulding can be a service-able and reasonably attractive object but, naturally, such a process was a crying invitation to shoddy and ugly design. Moreover, in order to get the powder to flow easily and quickly in the mould, and thus keep down the cost of manufacture, the fibre added was very short in length and its power to reinforce the resin was quite modest so that the ordinary commercial powder moulding was both weak and brittle.

The immediate result of the introduction of moulding powders was to extinguish the Birmingham brass trade. A secondary and rather ineffectual result was a wave of indignation on the part of the consumer both on account of the ugliness and the brittleness. I have been taken aside often enough and told 'They say they put sawdust into the stuff to make it cheaper?' One then had to explain that without the sawdust it would have been even worse and that, anyway, what could you expect when the trade price of small mouldings such as switch covers was about three shillings a gross? Latterly the design has improved a good deal, partly as a result of the competition from the much tougher thermoplastics such as Polythene and Nylon.

Powder mouldings can be made, in effect, by throwing a (weighed) handful of moulding powder into the hot mould and pressing the button which causes the press to close. However complicated the shape of the mould the powder flows like a liquid to fill it. This process is very valuable when we want to make elaborate little gadgets and especially for electrical devices such as wall plugs where the plastic has to flow around numerous brass pins and inserts. However, to get this flow it is necessary to use quite short fibres and, as we have said, the result is a material which is comparatively weak and brittle. One reason for this is that, with very short fibres, a crack in the resin which encounters a fibre has only to make a shortish detour to get around the head of the fibre and then proceed on its way.

Cellulose fibre laminates

Material of the highest strength needs to be reinforced with long fibres carefully packed together and for such systems the flow in the mould is very limited. These materials are therefore different in character from moulding powders. In the laminates developed in the 1920s cellulose paper or fabric was impregnated with a solution of phenolic resin, usually in alcohol, dried, and then the impregnated sheets were laid between the carefully trued, parallel, heated plattens of a hydraulic press where they were hardened under a pressure of about a ton/in^2 (15 MN/m^2).

The resulting material was expensive but of good quality and in some grades the strength and toughness were quite high. Phenolic resins are either black or dirty brown and so these boards were not used for decorative purposes. Originally most of the paper-based material was used for electrical insulation; the fabric-reinforced laminates, being very tough, were sold for machining into cams and gears and bearings. After the last War, melamine resins, which were colourless, came in. This enabled the surface to be made from a coloured and patterned paper, the bulk of the thickness being still of phenolic impregnated brown paper which is cheaper and stronger. As might have been predicted, material of this type has been very successful for table tops and panelling and it has played a big part in the Kitchen Revolution.

The decorative plastic sheets which are generally sold are comparatively weak and brittle but, since they are nearly always glued to some kind of rigid substrate, such as a wooden bench, this does not matter very much. It is not very easy to realize that before these materials came on the market no really satisfactory surface existed for table tops and an unthinkable number of woman-hours were spent in scrubbing wooden surfaces which, because of their porous nature, merely absorbed the dirt.

Although the cellulose in cellulose-phenolic laminates retains most of its affinity for water the worst exuberances of the cellulose can be restrained by drying the fibre before the material is mould-ed and then moulding and hardening the material in as dry a state as possible. When this is done the fibre is, as it were, clamped and restrained by the matrix and by the other fibres which cross it, so that, although all resin matrices are permeable to water vapour, the swelling is much reduced. Since the paper or fabric must be coated with resin at an early stage in the manufacturing process and the drying must be done immediately before pressing, the resin naturally gets dried as well as the cellulose. Now the ease with which a phenolic resin flows in a hot mould before it sets depends very sharply upon the amount of water present, so the dry resin needs a high pressure to get it to distribute itself uniformly and bond the material together properly. For this reason it is usually not possible to manufacture such materials with reasonable water resistance at pressures much below a ton per square inch. The total load which has to be applied by the press to a standard four foot by eight foot panel is therefore about five thousand tons, which is one of the reasons why the process needs a certain amount of capital.

The moisture resistance is also affected by the chemical details of the impregnating process. It is possible to reduce the moisture pick-up considerably by the right choice of resin and this is often done with materials for electrical purposes. Unfortunately good moisture resistance means blocking the hydroxyls in the cellulose and this makes it brittle and so unsuitable for mechanical applications. Immediately after the War I saw an aeroplane which the Germans had built from a paper-based phenolic material. To ensure toughness they had reduced the water resistance of the material as far as they dared – too far as it turned out. When I saw it, it had been in the open for about three months and was falling to bits.

A lot of work was done during the War in this country to develop cellulose-reinforced sheet as a potential substitute for aluminium sheet for covering aircraft. Retaining reasonable toughness, we managed to get the total moisture movement in the plane of the sheet down to 0·8 per cent and then covered parts of the surfaces of twelve service aircraft with the material by way of experiment. The plastic material never caused an accident but it did give a lot of trouble. The sheet was, of course, riveted to an aluminium framework which did not shrink and swell with it. In consequence, in the desert the plastic became so taut that cracks appeared along the line of rivets while, in a wet climate, and especially after melting snow, the sheet buckled and waved in an alarming way. Eventually we pulled the material off. In practice the dimensional movement of cellulose-reinforced materials will be about 1·0 per cent or about one inch in eight feet. This matches neither metal nor wood nor plywood and generally condemns the material for large-scale applications.

Nowadays the use of strong cellulose laminates is almost

confined to flat sheets which can be pressed between accurately trued plattens. When shaped shell mouldings are wanted two shaped steel dies are needed. These are expensive anyway (and nearly impossible to alter) but what makes shaped moulding peculiarly difficult is that the material has almost no flow during pressing. For this reason the clearance between the two halves of the mould must be very accurately maintained. If this is not done, then all the press load will be carried at the tight spots and the rest of the material will not get pressed at all. The difficulty and expense of all this is generally enough to frighten people off, especially as there are nowadays several better ways of getting shaped mouldings. Around 1940, however, this was not so and a few serious large mouldings were made in spite of the heavy tool cost. The best remembered of these was the standard fighter pilot's seat which was used in the Spitfire and elsewhere. This was a fairly large and elaborate structure, made by bolting several shaped mouldings together. It had to withstand loads in the region of a ton and never gave serious trouble. On the other hand the saving in weight and cost over a riveted metal seat was not very great.

Glass fibre materials

Modern reinforced plastics really date from the introduction of inorganic fibres about the end of the War, mainly, in the first instance, for radomes; that is, for the dome-like structures which house radar scanners and therefore have to be transparent to radar and thus must be made from electrically non-conducting materials. The most successful of these fibres so far is glass. Except that the composition of the glass is sometimes changed in detail, these fibres are identical with those pulled by Griffith over fifty years ago. The drawing process has been mechanized in that the glass is generally melted in an electrically heated platinum cistern in the bottom of which are usually either two or four hundred small teat-like holes. From each of these teats a fibre is pulled, the glass being hard and cold by the time it reaches the revolving drum beneath the furnace on which it is wound. Commercial fibres are usually about a third to half a thousandth of an inch thick, from five to ten microns. The pristine tensile strength of the fibre is perhaps 400,000 to 500,000 p.s.i. (3,000 MN/m²) though this gets reduced by subsequent handling. Because the new fibres tend to stick to each other and cause mutual weakening, the fibre is treated with an organic protective film applied between the drawing teat and the winding drum. This film enables the fibre to be handled during subsequent processes, such as weaving, with less damage than would otherwise occur. On impregnation the film combines with the bonding resin.

What happens after the fibre is drawn and reeled up on the winding drum depends upon the purpose for which the moulding is required. As we said earlier on it is necessary to get as much fibre as possible packed into the moulding, simply because the fibre is at least a hundred times stronger than the resin, so that, other things being equal, the strength of the finished material is proportional to the fibre content. The solids content of loosely packed individual fibre mats is very low indeed so that the fibre is seldom used in this form except for special purposes such as the so-called 'finishing sheet' which is often used to get a good surface on mouldings. Fibres pack best in parallel bundles, that is to say as threads or yarns. These yarns usually contain several hundred individual fibres. Because these fibres are continuous it is not necessary to put much twist into the yarn to hold it together.

Sometimes such glass yarn, after being impregnated with resin, is wound into cocoon-like structures intended for tanks and pipes and pressure vessels. For many high-grade applications however the yarn is woven into a cloth which looks not unlike an expensive white satin. During the days of clothes rationing and before glass cloth was widely familiar I had a large roll of such cloth stolen from the laboratory, no doubt to be made into underclothes. Since glass fibre is an irritant to the skin I watched the women employees, over a considerable period, to see if they scratched themselves. However, either they were all innocent, or else possessed of great self-restraint, for I never caught anybody.

Though the strength of laminates made from glass cloth is good, the cost is fairly high, not only because the woven cloth is expensive but also because cloth is not very suitable for automatic handling in making shaped mouldings. The greatest tonnage of glass fibre reinforcement is therefore handled as what is called 'chopped strand mat'. In this the yarns are chopped up into short lengths, usually of two or three inches. Most of this chopped strand goes into the manufacture of flat mats which are made by blowing the short lengths of yarn against a wire gauze with a little adhesive. This adhesive dries very quickly so that the mat can be removed from the gauze and handled like paper. For shaped mouldings suitable pieces of this mat are cut up and attached to the mould until the component is built up to the required size and shape.

When considerable numbers of shaped mouldings are to be made a blown preform technique may be used as this is almost a fully automatic process. The process is used for such things as helmets and typewriter cases. Instead of blowing the fibre on to a flat gauze it is blown and sucked by air on to a shaped gauze mould after which the lightly stuck together preform is automatically transferred to a heated steel die in which the main bonding resin is applied and cured under pressure.

Apart from its high strength, glass fibre has the great advantage of not swelling in water and for this reason, among others, it is not necessary to carry out the moulding operation under high pressure. This means that, if required, cheap moulds can be used, which can be altered about as the design is changed, and large hydraulic presses are unnecessary.

It is possible to bond glass fibre mouldings with phenolic resins but generally it is better to use the special resins, such as the polyesters and epoxies, which are developed and sold for the purpose. Many of these will set, not only with negligible pressure, but also at room temperature, after a catalyst has been added.

This has lead to what is rudely known in the trade as the 'bucket and brush technique'. These methods, which are popular with amateurs and with small firms, are almost identical with Egyptian papier mâché. Alternate layers of cold-setting resin and glass mat or fabric are laid up by hand over a simple plaster mould and left to set in their own good time. Given careful and conscientious methods there is not very much wrong with this construction, though it is expensive in labour if more than a few mouldings are needed. For very large mouldings, such as boats, it is the only practicable way of doing the job.

One of the objections is that no two mouldings are the same and that adequate inspection and supervision are almost impossible. Since it is difficult to predict the strength of any individual structure, the method has drawbacks for aircraft parts. For reliable mouldings the resins must be set in a dry, warm, controlled atmosphere and this may not be available in a back-street workshop. Many of the complaints about glass fibre in boats can be traced to the cold, damp shed of some local boat-builder. The larger firms are taking to using carefully heated (and expensive) building sheds and perhaps the most satisfactory role for the local builder or for the amateur is that of finishing off moulded hulls made by professionals with proper facilities.

For large mouldings like boats the cost of the mould is important because the number of orders is not usually very great. It is better to use a cheap mould and let the resin set slowly at room temperature. Furthermore one can put up with a lot of hand finishing on the moulded shell. When we come to such things as helmets and luggage the economic picture changes. In these cases it is usual to provide matched pairs of heated steel dies. A glass fibre 'preform' is dropped into the die and then a determined quantity of hot setting liquid resin is added at the last moment before the die closes. The setting rate is arranged so that the resin has just time to penetrate the fibres uniformly before it hardens and the moulding is ejected. Because of the smooth finish on the steel mould the moulding needs little hand labour in the final smoothing. The whole process of making the fibrous preform, laying it in the mould, impregnating it with resin and hardening the resin may be carried out in one large machine in a matter of seconds as against hours or days for the hand lay up process.

In the earlier reinforced materials small amounts of fibre were used to ameliorate the worst faults of a weak brittle matrix. Such materials could be correctly spoken of as 'reinforced'. In the newer materials the function of the matrix is simply to glue together a number of strong fibres and, as the weak constituent, we use as little of the matrix as we possibly can while still effectually bonding the fibres. Such systems would be more accurately called 'bonded fibre materials' – in fact, they are often called 'fibre composites' nowadays.

A serious study of the properties of these systems is a difficult and highly mathematical subject which has latterly become respectable and indeed fashionable in academic circles. Without going into any detail it may be said that the properties of a lot of fibres glued together are more or less what one would expect from engineering common sense.

It is usually difficult to get more than about fifty per cent by volume of fibres into any material. Glass fibres may be taken as having a tensile strength of about 300,000 p.s.i. (2,000 MN/m²) by the time they have been processed and a Young's modulus of 10×10^6 p.s.i. (70,000 MN/m²). A material, like a fishing rod, made from parallel glass fibres in resin will therefore have a tensile strength of about 150,000 p.s.i. (1,000 MN/m²) and a Young's modulus of 5×10^6 p.s.i. (35,000 MN/m²), since the resin contributes almost nothing to the stress situation, though of course it adds to the weight. Calculating from the simple law of mixtures, the specific gravity of the material will be 1·85, if all the air voids are filled as they should be. (The specific gravity of glass is usually 2·5 and that of resin 1·2.) We can therefore put down the following simple comparative table:

Material	S.G.	Tensile Strength (p.s.i.)	MN/m²	Strength for weight (p.s.i.)	MN/m²	Young's modulus (p.s.i. $\times 10^6$)	MN/m²	$\dfrac{E}{S.G.}$ (p.s.i. $\times 10^6$)	MN/m²
Parallel fibre-glass	1·85	150,000	1,000	81,000	550	5·0	35,000	2·7	19,000
Crossed fibre-glass (woven fabric)	1·85	75,000	500	40,000	280	2·5	17,000	1·85	9,500
Mild steel	7·8	60,000	400	7,700	50	30·0	210,000	3·85	27,000
High tensile steel	7·8	300,000	2,000	38,500	260	30·0	210,000	3·85	27,000

From which it will be seen that direct comparisons between steel and fibre-glass are not very simple. However, very roughly, fibre-glass is stronger than steel, especially for its weight, but it is not nearly so stiff, even when we take account of its much lower density. In this respect it is also worse than wood.

Like wood, comparisons depend to a certain extent upon how many directions one wants to be strong in. Naturally such materials show up best when all the fibres and all the strength are put in one direction, though the engineering applications for unidirectional materials of this kind are fairly limited. When equal numbers of fibres are crossed at right angles we get a material like plywood, with half the strength of a unidirectional material at 0° and at 90° but rather less at 45°. This can be achieved with woven glass-fibre cloths.

Theory says that, if we want truly uniform properties in all the directions in a fibrous sheet material, then there are several arrangements of fibres which will give it and they will all achieve strengths and stiffnesses equal to one-third of that of a unidirectional system. Experiment shows the theory to be perfectly right here, provided of course that the resin contents are comparable. However, the case which generally occurs in practice with fibre-glass is that of chopped strand mat. With this reinforcement it is seldom possible to attain fibre contents as high as fifty per cent (because the fibres do not pack so well) and so we have generally to reckon on rather less than a third of the unidirectional strength in real life. However, chopped strand mat is usually used for comparatively cheap jobs where the utmost strength may not be needed. Even so, chopped strand mat will generally beat mild steel, strength for weight. It is in stiffness that reinforced plastics in general, and fibre-glass in particular, cannot compete with metals and wood. This is one of the main difficulties about using the material in largish structures such as boats and car bodies and it quite rules it out for the main structure of conventional aeroplanes at present. It is true that one can stiffen up car bodies, for instance, by putting in inserts of steel tubes, but then much of the attractiveness of the job has gone.

Metals, of course, are nearly isotropic, that is to say they have almost equal properties in all three dimensions. This is very useful if one wants to do the sort of things one can do particularly well with metal, such as making engine crankshafts, but it is not particularly useful for the shells and panels which one usually wants to make out of fibrous plastics. This is just as well because it is almost impossible to get isotropic properties in practice in a fibrous material because it is difficult to get fibres to pack tightly and to point in three directions at once. Even a haystack is apt to be a layered structure. Theory indicates that the strength of a three dimensionally random arrangement of fibres would be one-sixth of that of an all-parallel system, so it is really not worth trying anyway.

The work of fracture mechanism in fibre composites like fibre-glass is not without interest. As we have said, the work of fracture of glass fibre itself is no better than that of any other kind of glass, which is to say, abysmally bad. The resin is, perhaps, a little tougher, but not very much so. When the two are put together however the resulting composite is reasonably tough – tough enough, at any rate, for things like boats and crash-helmets.

The way in which this works is a little complicated. If we sup-pose a crack to be proceeding through the resin it will very soon encounter a fibre; if the material has been properly made – that is to say, if there is not too much and not too little adhesion between the fibre and the matrix – then the fibre will not be broken at that point but the material will crack at the interface between the glass and the resin for the reasons which we discussed in Chapter 5. This crack will spread along the fibre (this is called 'crack-back') so that the fibre becomes detached from the matrix, often for a considerable distance (on the left of Plate 11). Further-more, for reasons which have been elucidated by Richard Chap-lin, the crack in the resin is very apt to fork wherever it meets a fibre so that the number of cracks in the resin is greatly multi-plied. We can generally see this process quite easily whenever a fibre-glass article has suffered from a blow because the material in that region, though not broken, usually turns white. This whiteness is due to the reflection of light from the surfaces of the many internal cracks. Material in this condition is not much weaker than it was before, although it has already absorbed a good deal of energy, simply in providing all those internal surfaces.

Before the composite can actually break, all the reinforcing fibres must, of course be broken. Since the strength of glass is very variable, the fibres do not break in any one plane in the composite but rather, fractures occur in a scattered or random manner throughout a considerable volume of material. The final phase of fracture therefore involves pulling all these fibres out of their holes or burrows. In order to do so a great deal of friction has to be overcome and this is where the bulk of the work of fracture of composites of this type comes from.

The whole process had been reduced to algebra by Kelly and others and it is found that the experimental works of fracture correspond pretty nearly to the theoretical maximum and so not much further improvement is to be looked for in composite materials of conventional design. This work of fracture is usually around 10^3 J/m^2 which is distinctly on the low side as compared with, say, mild steel. This consideration becomes important when we come to consider such matters as the safety of car bodies made from steel and from fibre-glass. Using his helical fibre work of fracture mechanism George Jeronimidis has recently been able to make glass-fibre composites with works of fracture in the region of 10^5 J/m^2 – roughly a hundredfold advance – and this may perhaps afford a break-through in this respect.

With all their drawbacks materials like fibre-glass are slowly spreading and becoming more widely used as they become more widely understood and appreciated. The cost per pound of the raw materials is something like fifty pence and this seems a great deal compared with steel at about 5p and aluminium at perhaps 30p. However the proof of the pudding is in the fabricating. The cost of making complicated shapes in plastics is usually so much less than it is in metal that the finished structure may well be cheaper. However, before this can happen it is usually necessary drastically to redesign the product and this may meet with resistance.

Very much depends upon the type of product. Nobody would try to make, say, a petrol engine out of plastics. But then metals are usually inefficient for light thin shells. There is quite a good argument for building the hulls of larger boats in steel, at least if one isn't in a hurry and doesn't mind the weight, but steel is hopelessly inefficient for the smaller hulls because the plates become so thin that, even if we can put up with the buckling, denting, oilcanning or whatever one likes to call it, a few months' corrosion will eat right through. In this field fibre-glass seems to have established itself very firmly and the costs are certainly competitive.

There have been many improvements in motor cars during the last generation, but in my opinion, the pressed steel body is not one of them. It is enormously heavy and its weight puts up the petrol consumption and reduces the performance, it requires a great deal of sound-proofing, and, worst of all, it starts to rust soon after one first gets the car home and body corrosion, rather than mechanical wear, is probably the reason why the majority of cars are eventually scrapped.

The use of fibre-glass in the bodies of mass produced cars is generally inhibited by three things. It is still marginally more expensive than steel in large-scale production and also it is difficult to get quite that showroom gloss on the surface finish which the motor trade seems to think that the public want. Furthermore – and more importantly – the crash protection afforded by a steel body is much better than that of a glass-fibre one – for the reasons we have just discussed. Yet nearly all the numerous non-mass production cars seem to have fibre-glass bodies. Indeed this is probably the factor which enables these cars to keep going economically because the sales are too small to justify die-pressed steel bodies while the old-fashioned 'coach-built' body would be too expensive anyway. By using a fibre-glass body the specialist car builder can turn out a car of roughly half the weight and thus get a flying start in performance at an extra cost which is often only about a hundred pounds. In spite of its disadvantages the world production of glass-fibre mouldings is said to have reached nearly a million tons per annum and to be increasing rapidly. This compares with a world production of about four and a half million tons for aluminium and its alloys. The expansion of the glass-fibre industry is likely to be limited, in the end, by the relatively low stiffness of the material.

High stiffness composites and carbon fibres

Over the last fifteen or twenty years various Governments have spent a great deal of money on the development of high-stiffness composite materials, some of which have received publicity out of all proportion to their actual usage. The main incentive for this work has been the requirement for weight-saving in critical structures and especially in aero-space, for it has been calculated that it would be practicable to treble the payload of long-range commerical aircraft by the use of improved structural materials and there are comparable advantages with space-travel and in military applications.

Before going into details it is worth reviewing the thinking which lies behind the development of advanced materials for applications of this sort. Presumably we can do very little to change the properties of natural wood but we might perhaps change the properties of a metal, such as aluminium, or else we might substitute a new and better metal. At first sight it would seem that all we have to do is to increase the strength of our metal – after all aircraft parts are designed as near to the bone as possible, that is as near to the breaking stress as seems safe. As we shall see it is not unduly difficult to strengthen a metal and, if the material were stronger then the part could presumably be made thinner and therefore lighter. To a limited extent this is clearly true – but only to a rather limited extent.

It will be remembered that, although we can now modify the strength and toughness of solids very considerably we have no real control over their stiffness. The Young's modulus depends solely upon the chemical nature of the solid and cannot, as a rule, be changed by tinkering with it. If we want a different modulus then we must change to a different substance. Thus, if we increase the strength of any solid, such as a metal, we do so by increasing its elastic breaking strain and, in order to make use of its higher strength we must operate it at higher strains. That is to say that the deflections of the structure as a whole will be higher; if we have put up the stress a lot, so as to save a useful amount of weight, then the deflections will be much higher. The consequences of this kind of action can be seen in the picture of the bent aeroplane in Chapter 2 and clearly such shapes are not acceptable.

Another reason why we need stiffness is that much of the structure of an aircraft, for instance, is in compression and moreover the compression parts are usually struts and plates which are thin in proportion to their length. Members of this sort fail under compression, not by direct crushing, but by elastic buckling and the cause of such failures is not lack of strength but lack of stiffness. This is called 'Euler collapse'.

Yet again, many parts of an aeroplane are liable to break, not by simple loading in one direction, but by what is known as flutter. That is by flapping violently in the airstream like a flag. This is guarded against by increasing the stiffness of the parts, not by increasing their strength.

For these reasons we find that if we simply increase the strength of a material we shall soon run out of Young's modulus and it is therefore as important to increase the Young's modulus as it is to increase the strength. Now when we are dealing with structures of minimum weight, as we are in things like aeroplanes, we are not so much interested in the actual or absolute properties of our materials as in their specific properties; that is, how much strength or stiffness we are getting for a given weight. Specific figures are obtained by dividing the actual values by the specific gravity or density of the material. It is salutary to look at the Young's moduli of the common engineering materials in this light. The figures are given in Table 1.

TABLE 1

Young's moduli of orthodox structural materials

Material	Specific gravity grams/c.c.	E		E/S.G.	
		p.s.i. $\times 10^6$	MN/m²	p.s.i. $\times 10^6$	MN/m²
Molybdenum	10·5	40·0	270,000	3·9	25,000
Iron and steel	7·8	30·0	210,000	3·8	25,000
Titanium	4·5	17·0	120,000	3·9	25,000
Aluminium	2·7	10·5	73,000	3·9	25,000
Common glasses	2·5	10·0	70,000	4·0	26,000
Magnesium	1·7	6·0	42,000	3·7	24,000
Wood – spruce parallel to grain	0·5	1·9	13,000	3·8	25,000

It is remarkable that the specific Young's moduli of all these materials should be almost the same and, although this is more likely to be due to chance than to any deep philosophical reason it does make things a bit awkward for engineers and materials men. It also accounts for the fact that, for a great many years, steel and aluminium alloys and magnesium and wood (and latterly titanium) have competed with each other in the aircraft industry on roughly level terms. If you design airframes for the same aircraft in any or all of these materials they are apt to come out at much the same weight.

This is tantamount to saying that, if we want to make any large or dramatic advance in materials for this kind of purpose, then we shall have to reject all the common engineering solids about which we have accumulated experience. What can we do about it? What chemical entities are there which have higher moduli? The answer to this is – not an enormous number but quite a lot. Some of the better ones are listed in Table 2.

TABLE 2

Some high specific modulus materials

Material	S.G.	E p.s.i. $\times 10^6$	E MN/m²	E/S.G. p.s.i. $\times 10^6$	E/S.G. MN/m²	Melting point °C
Aluminium nitride (AlN)	3·3	50	340,000	15	103,000	2,450
Alumina (Al₂O₃)	4·0	55	380,000	14	95,000	2,020
Boron (B)	2·3	60	410,000	26	180,000	2,300
Beryllia (BeO)	3·0	55	380,000	18	130,000	2,530
Beryllium (Be)	1·8	44	300,000	24	170,000	1,350
Carbon whiskers (C)	2·3	110	750,000	48	330,000	3,500
Magnesia (MgO)	3·6	41	280,000	11	78,000	2,800
Silicon (Si)	2·4	23	160,000	10	66,000	1,400
Silicon carbide (SiC)	3·2	75	510,000	23	160,000	2,600
Silicon nitride (Si₃N₄)	3·2	55	380,000	17	120,000	1,900
Titanium nitride (TiN)	5·4	50	340,000	9	63,000	2,950

In some ways Table 2 is a cheering document because it shows that there are in existence solids which have specific Young's moduli which are, very roughly, ten times as high as those of the traditional engineering materials and so they hold out the possibility of a rather spectacular 'break-through'. In other ways the list is a daunting one. All these materials are normally very weak and brittle, they can usually only be made at all at very high temperatures and some of them are toxic.

The only metal in the list is beryllium. Now beryllium can be dangerously poisonous (for instance it may be fatal to get it into a cut), but supposing for the moment that we discount the toxicity, can it be made tough and strong? The answer seems to be that in certain cases it can be made fairly strong but that it is very difficult to make it reliably tough. This is for the fundamental reason that, at normal temperatures, dislocations are only mobile in four planes in the beryllium crystal and, as we shall see in Chapter 9, we really need dislocation mobility in five planes if the crystal is to be immune from cracks attacking it from any angle. Although this characteristic of beryllium was predicted by the theoretical crystallographers quite a long time ago a great deal of money and effort has been wasted by the British and American Governments in trying to make ductile beryllium.

Though one cannot make beryllium reliably ductile one might perhaps be able to make it tough by making use of the composite approach. Working on these lines the Americans have experimented with matrices of magnesium or of magnesium-lithium alloys, reinforced by plates or particles of beryllium. So far, this has been only moderately successful, partly because magnesium – and to an even greater extent magnesium-lithium alloys – corrodes very rapidly. Another approach remains to be tried. One might be able to toughen beryllium by putting into it a small percentage of fibres, like wood-pulp in ice. However, I do not think that this last method has actually been tried or, if it has, I have not seen any results. If any of these experiments proved practicable I suppose that one might be able to arrange for some sort of protection from toxic hazards; yet a further objection lies in the fact that beryllium seems to be an unavoidably expensive metal, partly because of the precautions which have to be taken.

If we cannot civilize beryllium what else might one do? Clearly one might try to make a composite material based upon fibres which were much stiffer than ordinary glass fibres. Most conventional glasses have, very roughly, the same Young's modulus but again, a great deal of time and money has been spent by Governments in trying to develop high-modulus glasses suitable for fibre-drawing. It turns out that the only way to get a large increase in stiffness is to incorporate beryllium oxide, beryllia, in the glass. This is even more toxic than beryllium metal and the project has had to be abandoned for this reason.

One is therefore driven to try to make strong fibres from one of the other substances in Table 2. In the matter of dislocation mobility we require all or nothing. If dislocations are really mobile then we can make a reliably ductile material, like a metal, and do not need to bother with fibres. However, if we do make fibres, then the thing is to avoid any dislocation movement at all. Such movement does no good and it may do a lot of harm by weakening the fibre. For this reason beryllium and magnesia are not very suitable as fibre-forming materials.

Most of the other solids in Table 2 however are predominantly covalent bonded so that the dislocations remain immobile at normal temperatures – which is just what we want. In nearly every other respect however, most of these substances are what the materials man calls 'perfect stinkers'. Generally speaking they can only be formed with considerable difficulty and at high temperatures. If the substances themselves are not toxic then one or more of the ingredients needed to make them probably is. Moreover some compound needed in the process is almost certain to attack the walls of the apparatus.

It was pointed out in Chapter 4 that the problem about strength is not so much to explain why materials are strong as to explain why they are weak. In other words all solids are 'naturally' strong unless they are weakened by defects which are nearly all of a physical nature. It will be recalled that, in the case of brittle solids in which the dislocations are immobile, the weakening defects are small geometrical irregularities which set up stress concentrations. If the interior of the solid is reasonably homogeneous, then the dangerous defects are usually surface irregularities. Most of the covalent solids in Table 2 normally occur either as powders or else as irregular lumps. The task is to get these substances into the form of threads which are homogeneous within and smooth without.

This is what several laboratories, in England and abroad, have been working on for a number of years and there are now a variety of processes which will produce very stiff, strong fibres but none of these processes is easy or very cheap. For one thing the manufacturing temperatures needed are usually somewhere in the range between 1,000 °C and 3,500° C and at present the fibres have usually to be treated at high temperatures for hours or days. The engineering problems of getting furnaces to operate under corrosive conditions while maintaining reasonable chemical purity in the reaction chambers are severe. Moreover the amount of energy used per kilogram of fibre produced is very high.

These 'superfibres' usually fall into one of two broad categories, whiskers and continuous fibres.

Whiskers

These are not different in kind from the whiskers of water-soluble substances which we talked about in Chapter 3. It is however enormously more difficult in practice to grow whiskers from the covalent ceramic compounds than it is from, say, hydroquinone. There are a number of processes and all of them are scientifically very complicated. Processes intended to produce high quality whiskers are usually vapour-phase ones, that is to say the ingredients of the reaction are handled as vapours in high temperature furnaces. There is no difficulty in getting gases such as nitrogen or oxygen into the reaction zone but elements such as silicon, carbon or aluminium cannot be transported satisfactorily as elements and it is necessary to use what is known as a 'transport species'. These may be such compounds as SiO, $SiCl_2$, CH_2, AlO and so on. The object is to produce strong, smooth whiskers in quantity and as quickly and cheaply as possible. Unfortunately, when the reaction is speeded up the quality of the whiskers grown is liable to be reduced and the whole problem is very difficult. The extremely complicated chemical reactions which occur are analysed by computers.

So far practical whisker farming has been attended by only rather moderate success. Silicon carbide whiskers have been grown in kilogram quantities in England, in Switzerland and in Japan and can be bought commercially. It is also possible to buy alumina (sapphire) whiskers in smallish quantities from America.

At present the demand for whiskers is limited because the price is high and the price is high because the demand is limited – moreover the fashion has been for continuous filaments such as carbon fibres. There have been fairly persistent rumours that the Russians have been growing diamond whiskers; however, since I was nearly put in prison for asking questions on the subject behind the Iron Curtain, it seems safe to assume that no official information is available.

Continuous fibres

As reinforcing fibres, whiskers have several potential advantages. In the long run they seem to offer the possibility of cheap large-scale production (though this has not yet been achieved), as short fibres they are much better suited than long ones to cheap fabrication by modifications of paper-making methods and, moreover, they are generally very much stronger than continuous fibres since, because of the growth processes which are used, their surfaces are almost perfectly smooth. Most good quality 'commercial' whiskers reach a mean elastic breaking strain of about 2 per cent, which corresponds to a very high stress. The importance of this lies not so much in the possibility of making materials of very high tensile strength – as we have seen this is not generally the most important problem – but rather that, exploiting modern composite theory, this strength can be employed to increase the work of fracture of the composite. As we shall see, it is lack of work of fracture which is generally the fatal handicap in advanced continuous fibre composites of the carbon-fibre type.

Having burnt their fingers once over the brittleness of beryllium one might perhaps have supposed that Governments would have learnt to pay some attention to theoretical predictions about work of fracture but this does not appear to be the case and, of course, continuous fibres do have some substantial advantages. Thus most Governments have chosen to put their major effort into continuous fibre developments. The properties of the principal continuous fibres are given in Table 3.

The first of these fibres to be developed was boron – an entirely American achievement – dating from about 1960. In the preferred process a thin (0·013 mm) tungsten wire is drawn through a reaction chamber which is filled with either a mixture of boron trichloride (BCl_3) and hydrogen or else with a mixture of boron hydride (B_2H_6) and chlorine. The tungsten wire is heated electrically, like a lamp filament, to something over 1,000°C when a fairly thick (0·05 mm) layer of boron is deposited on its surface. What is wound up on a drum at the end of the process is therefore a thickish (0·1 mm) fibre consisting of a thin tungsten core surrounded by a thick skin of boron. Both boron trichloride and boron hydride are beastly substances but the process works quite reliably although it is both slow and expensive. This invention was greeted with great publicity and was described by a five-star general as 'the greatest technological break-through for three thousand years'. In fact composites based on boron fibres have been developed with both resin and aluminium alloy matrices and they have been used in service aircraft, in helicopters and in space-travel. The cases of very large rockets have also been made by winding tubes from boron fibre and resin and it is claimed by sober and responsible Americans that the development costs have been more than paid off by specialized uses of this sort. In any case the development cost of boron fibres – which might have been one or two hundred million dollars – would be a drop in the ocean compared with the American aero-space budget.

Although all this seems to be true and the material has many technical virtues, boron composites have not caught on for commercial purposes because the process of making the fibres is inherently expensive (about $700 per Kg.) and, moreover, boron

TABLE 3

Some modern high-stiffness reinforcing fibres

Fibre	S.G.	Young's modulus		Tensile strength	
		p.s.i. × 10^6	MN/m²	p.s.i.	MN/m²
Boron	2·5	58	400,000	450,000	3,100
Carbon	2·2	60	410,000	300,000	2,000
Kevlar 49	1·45	19	130,000	400,000	2,700

fibres do not easily lend themselves to cheap fabrication processes.

Carbon fibres of a rather fragile kind date back to the end of the last century when Edison made electric lamp filaments by heating bamboo. However about 1964 Bill Watt at Farnborough was able to carburize polyacrylonitrile fibres – which are the basis of the dress-fabric 'Courtelle' – under rather special conditions so as to produce a fibre which combined a very high modulus with substantial tensile strength. Like boron, this fibre has suffered from a great deal of irresponsible Government publicity.

Although frequently described by the newspapers and on television as a wonderfibre of 'exceptional strength', carbon fibres are not, in fact, particularly strong; if anything they are a little weaker than glass fibres. They are however, for their weight, something like eight times as stiff as either glass or the normal engineering metals. As might be expected the resin-fibre composites made from carbon fibres are very stiff but not especially strong in tension. They are also, at present, rather inconveniently weak in compression. While it might be possible, in theory, to put up the compressive strength of carbon fibre composites by using a metal matrix, this does not usually work in practice because of the chemical reactions which occur between the carbon and the metal. In fact most attempts to manufacture carbon fibre-metal materials turn out merely to be very expensive ways of making acetylene gas (C_2H_2).

However, for many purposes where weight saving is important but where the strength requirements are not too critical – such as artificial limbs, golf club shafts or the stiffening of car bodies – carbon fibre composites have been very successful. When we turn to more exacting applications, like aircraft parts, the trouble is generally lack of sufficient toughness. When the composite is made in the conventional way the work of fracture which is actually achieved is not far short of the calculated theoretical limit – but unfortunately this limit is, in practice, not sufficiently high. This was at the root of the trouble with the carbon fibre fan blades in the RB211 engine which contributed to the Rolls Royce debacle a few years ago. These blades were unable to withstand the impact of the small birds which would have been sucked into them from time to time and, in the event, nothing could be done about this except to replace the plastic blades by titanium ones at the expense of the performance of the engine.

In the light of modern knowledge two ways of increasing the work of fracture of such materials suggest themselves. One way is to increase the strength of the fibre; if this can be done then the theory of the work of fracture of composites shows that the toughness can be increased. Although Bill Watt has been able to make considerably stronger carbon fibres on a small scale in the laboratory, it has not so far been practicable to put these fibres into production. The other way would be to use George Jeronimidis' timber mechanism, giving to the fibres a helical geometry. Working in the laboratory with carbon fibres and epoxy resins George has recently been able to make very dramatic improvements along these lines. It will be interesting to see what comes of this.

The last fibre on the list in Table 3 – Kevlar 49 – is something different, both because it is made from an organic polymer (with no nonsense about carburizing, which comes expensive) and also because it has been developed by private enterprise and not by Government defence laboratories. Messrs Du Pont, in America, have been quietly working away, to my knowledge, for at least twenty years on the development of high-stiffness fibres. They have explored the field very thoroughly and must be presumed to know what they are doing. Essentially, this material is made up from benzene rings joined together without folding, much as the sugar rings are joined in cellullose; in fact, both the density and the modulus are closely similar to those of a high-grade cellulose, such as flax. The strength however is about four times as high as that of the best flax and, of course, the material is virtually immune both to moisture and to rot. Weight for weight, the stiffness of Kevlar is not quite as good as that of boron or carbon but it is not so very much worse and the fibre may well prove to be both cheaper and more practical.

Reinforced concrete

Although the reinforced concrete people and the plastics people have been rather slow in communicating, the ideas involved are obviously parallel and it seems worth ending this chapter with a tailpiece on reinforced concrete. The origins of both materials seem to lie in the dim past and the differences are mostly in the matter of scale, the reinforcement being much coarser in concrete than in plastics. The Babylonians used reeds to reinforce structures of dried mud and various forms of 'wattle and daub' have been used all over the world. The Essex village in which I am writing this chapter is mostly built of mud or plaster over wattles, i.e. mud huts.

The use of iron as a reinforcement seems to be specifically Greek. As we said in Chapter 2, in normal masonry everything has to be kept in compression because masonry is not usually able to withstand any appreciable tensile stresses and so this condition leads normally to arches and domes which enable large openings to be bridged without the use of tensile stresses. However, although they were well aware of it, the Greeks seem to have rejected the arch, at least for formal architecture, very possibly on aesthetic grounds. In fact, in spite of John Keats, the Greeks were in some ways the least functional of people, especially in their architecture, which grew out of wooden architecture. The Parthenon and all other Doric temples are an exact copy in marble of wooden buildings, down to imitations in marble of the pegs which held the wooden beams together. As the result is dazzlingly successful and most of our own buildings are hideous we are in no position to jeer at the Greeks about this.

Wooden architecture is essentially an architecture of beams because wood is available in long pieces and has good tensile strength; Greek architecture was thus an affair of beams and columns. This is beautifully illustrated by American 'Colonial' architecture which, having timber in plenty, naturally reverted to classical styles with grace and success in wood. One cannot be 'Gothic' in wood because Gothic is a compression architecture originating in stone arches. Although marble is rather better than other stones in tensile strength it is really too weak and variable to make reliable beams of any length. In the early Doric stone temples this weakness was counteracted by keeping the spans of the beams short and by using very wide capitals at the heads of the columns. Even in the Parthenon (begun in 447 B.C.) the free span of most of the beams was kept down to about eight feet, though it looks more. However, when Mnesicles came to build the great gateway to the Acropolis, the Propylaea (begun 437 B.C.), both the architectural proportions and the need for ceremonial processions to enter called for much wider spans which in fact vary between thirteen and twenty feet. To deal with the tensile stresses Mnesicles caused iron rods about six feet long to be concealed and cemented within grooves in the marble. Thus we have iron reinforcing marble so as to make it behave like wood.

However, Mnesicles was not being particularly advanced because the Greek colonists at Akragas (Agrigentum) in Sicily were using iron reinforcing members fifteen feet long and of five-by twelve-inch cross-section as early as 470 B.C. How such large forgings were made is a mystery and it suggests that the Greeks would have had no technological difficulty in making steam engines and other heavy machinery if they had given their minds to it and if it had occurred to them to do so.*

As we have said, Gothic churches fell down if a tension load made its appearance, which it did quite frequently; the cure or palliative was to provide buttresses, pushing in the desired direction. Much the same thing applied to late classical and Romanesque architecture and although the outward thrust of the dome of St Sophia, in Constantinople (begun 532 A.D.) is countered by the inward thrust of the half domes on which it rests, the arches are tied across the base by iron tension rods.

The domes of St Peter's and St Paul's Cathedrals stand upon drums or cylindrical towers and there is no possibility of taking the outward thrust by means of subsidiary domes or buttresses which would have completely spoiled the effect of the designs with their isolated cupolas. As is well known the problem was solved in both cases by taking the thrust in a circular tension

* But of course the supply of fuel in the Ancient World would have been a difficulty.

chain which is embedded in the masonry around the bottom of the domes.

A more general approach was initiated by a Frenchman called Soufflot (1713–81) who tried to increase the tensile strength of masonry by burying a number of iron rods in the joints between the stonework. However, the wet got in along the mortar and rusted the iron, the expansion of the corrosion products then crumbled the masonry. The great I. K. Brunel (1806–59) later tried to do much the same thing, putting hoop-iron (that is the thin iron strip used for barrel hoops) between the joints of his brickwork with exactly similar expensive results.

Apparently it fell to three people to discover, almost simultaneously, that iron reinforcement in Portland cement did not rust sufficiently to cause damage. A French gardener, Joseph Monier (1823–1906), made flowerpots, or rather large tubs for orange-trees, by embedding a mesh of thin iron rods in concrete in 1849. These tubs were successful and attracted attention. An Englishman, W. B. Wilkinson, seeking a use for old mining ropes, of which a large quantity was available, made reinforced beams for use in building (much like the Greeks), putting the wires on the tension side of the beams. Finally, a French engineer, J. L. Lambot, exhibited in 1855 a rowing boat made of concrete reinforced with iron bars, presumably the first of a long line of not very successful concrete ships. Lambot patented, a little late in the day it would seem, the combined use of iron and cement in building.

Iron reinforcement does enable cement to carry tension loads quite successfully but the tensile breaking strain of concrete is low so that the concrete cracks long before the iron is seriously strained, and if any serious tensile load is put upon the combined system an elaborate pattern of cracks appears in the concrete. If these cracks are small they let the water in, if they are large the concrete may fall out piecemeal. To avoid this the best thing to do is to put the concrete permanently into compression by putting the steel reinforcement permanently into tension. In one form or another this arrangement, which is known as prestressed concrete, began to come in about 1890 but although it was quite successful it has been slow in catching on and it has only been applied seriously and on a large scale relatively recently. The use of pre-stressed concrete enables much more efficient and highly stressed structures to be built than were possible using concrete with ordinary reinforcement. One might say 'Why not make it all of steel?' In fact there is a considerable economy in the weight of steel used, mainly due to the fact that the steel has not to take the compression, and is, anyway, stabilized against buckling. Also, because the steel is protected from rust by the concrete, the maintenance of the structure may be nearly eliminated.

In the traditional 'reinforced concrete' the reinforcement – which usually takes the form of steel rods – is on quite a gross scale. However, during the last few years modern composite and fracture mechanics theory has been applied to the problem. This has resulted in the development of concretes with a fine-scale reinforcement. These materials are much more like the 'reinforced plastics' which we discussed earlier in this chapter. The reinforcement is generally thin steel wire but is sometimes a special sort of glass fibre. Though the potential for such materials in building construction and civil engineering seems to be large, there are still a number of difficulties to be overcome and at the time of writing, their actual usage is not very great.

Part Three

The metallic tradition

Chapter 9   Ductility in metals

or the intimate life of the dislocation

> *'We will now consider iron, the most precious and at the same time the worst metal for mankind. By its help we cleave the earth, establish tree-nurseries, fell trees, remove the useless parts from vines and force them to rejuvenate annually, build houses, hew stone and so forth. But this metal serves also for war, murder and robbery; and not only at close quarters, man to man but also by projection and flight; for it can be hurled either by ballistic machines, or by the strength of human arms or even in the form of arrows. And this I hold to be the most blameworthy product of the human mind. In order that death may reach men the more speedily, we attach wings to it; we deck iron with feathers and thus the fault is not nature's, but ours. A few examples prove that iron could in fact be an innocent metal. Thus in the alliances which Porsenna established with the Roman people after the expulsion of the kings, it was established that iron should be used for no purpose except agriculture.'*
>
> Pliny, *Natural History.*

As we said in Chapters 4 and 5, toughness, work of fracture and crack-stopping are very essential qualities in any useful strong material. To recap, 'toughness' implies, perhaps a little vaguely, a general resistance to the propagation of cracks; rather more precisely, 'work of fracture' is a measure of the energy which has to be consumed in propagating fracture through the material and thus of the length of a critical Griffith crack; 'crack-stopping' implies that the stress at a crack tip is reduced, usually by diminishing the stress concentration, to such an extent that the crack cannot proceed, even if the energy balance is in its favour. As we have seen, both natural and artificial non-metals rely upon a series of characteristic and highly ingenious mechanisms to provide these qualities. Metals achieve the same effects by a totally different method, basically by the dislocation mechanism – which scarcely occurs in nature. One effect of this mechanism is that many metals are 'ductile'.

If a metal possesses adequate ductility not only is the work of fracture high – which of course is a good thing – but the shape of its stress-strain curve is modified in such a way as to ensure large departures from Hooke's law. When this is the case one of the effects is that stress-concentrations are often very greatly and usefully diminished; in other words ductility tends to stop cracks. It will be realized that all calculations and assumptions about dangerous concentrations of stress assume that Hooke's law is obeyed. We speak of concentrations of stress but what the mathematical calculations really supply us with are concentrations of strain. Thus, if we calculate that the material immediately at the tip of a crack in a loaded structure is strained or stretched by 200 times more than the average for the structure as a whole then we assume that the local stress is also 200 times as high and we say that there is a stress concentration of 200. However, if Hooke's law is not obeyed by the local material at the tip of the crack then this is no longer true.

This is the classical metallurgist's way of crack-stopping but, since the local increase of strain at a crack tip is usually many hundredfold, it is of no use invoking minor departures from Hooke's law, such as might be caused by the shape of the inter-atomic force curve (Chapter 2). What is needed is a really whole-hearted lack of elasticity which is just what ductile metals provide. However, there is more than one way of departing from Hooke's law and it may be worth glancing at a mechanism which will *not* provide toughness.

Why viscous materials are no good

Viscous liquids will strain under a constant stress by any amount, if given sufficient time. In other words they flow. Like a quagmire, if you go on pressing, a viscous liquid will yield although it will resist a sudden load. All liquids are viscous but some are more viscous than others. The most viscous liquids are hard to tell from solids. In this class are pitch and tar, toffee and the baser sorts of plastics.

We have already remarked that a blow from a poker will shatter toffee which has resisted more adult and slower attacks and the same is true of pitch and the more deplorable plastics. These materials, given time, are very tough indeed because they can flow sufficiently at the crack tip to relieve the stress concentration. As structural materials, however, they combine the worst of both worlds. If they are subject to a sustained load they will flow in bulk and slowly run away from their responsibilities. Under a sudden load they are unable to yield in time and behave like solid glass. Once the crack is running it soon builds up to a speed to which the flow mechanism cannot possibly respond and so the material shatters.

Materials like wood and reinforced plastics creep a little, that is to say they behave in a slightly viscous way at high stresses and this, of course, is a bad thing. Quite apart from this creep however they are to a small extent non-Hookean so that the stress-strain curves generally look something like Figure 1. The departure from Hooke's law, though, is far too small to cause a useful

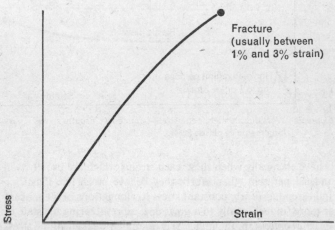

Fracture
(usually between
1% and 3% strain)

Stress

Strain

Figure 1. Stress-strain curve for typical non-metallic engineering material such as wood or fibre-glass. The departure from Hooke's law is not in general due to the shape of the interatomic force curve but is more often due to small creep effects.

reduction in stress-concentrations and these substances have to depend on weak interfaces to stop cracks.

Crack-stopping by dislocations – stress corrosion

At its best the dislocation mechanism provides a very desirable combination of elasticity at low strains with rapid and extensive yielding at high strains. A typical stress-strain curve for a ductile metal is shown diagrammatically in Figure 2. Such metals cease to

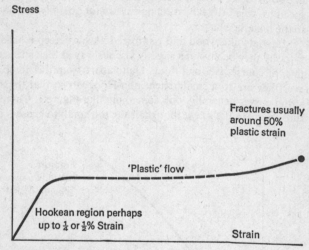

Stress

Fractures usually
around 50%
plastic strain

'Plastic' flow

Hookean region perhaps
up to $\frac{1}{4}$ or $\frac{1}{2}$% Strain

Strain

Figure 2.   With ductile metals the stress-strain curve usually shows a very long region of plastic flow.

behave elastically when they reach strains which are usually well under 1 per cent; thenceforth they behave much like Plasticine and extend at nearly constant stress to elongations of 50 per cent or more (in fact locally to a great deal more). During the state of 'plastic' extension, the material is not much weakened, the stress does not rise with increasing strain, but on the other hand the metal is not seriously damaged. The mean working strains deliberately put upon engineering structures seldom exceed about

0·1 per cent and, as the metal may be able to yield locally by 100 per cent or more, local concentrations of strain at crack tips in the region of perhaps 1000 to 1 can be accommodated.

Figure 3 shows that on either side of the actual tip of a crack there are small local regions of very high shear – shear stress

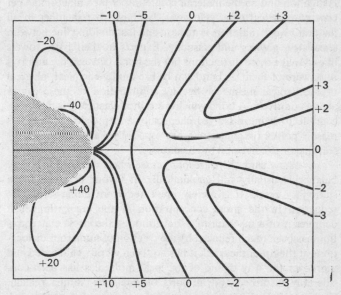

Figure 3. In addition to the concentration of tensile stress at the tip of a crack in a stressed material there are also concentrations of shearing stress which exist on either side of the crack tip. In a ductile metal these shear stress concentrations are able to nucleate hundreds of dislocations and thus help to relieve the stress concentration. This is a diagram of shear stress around a crack tip. Shear stress has a sign – positive or negative – either kind can nucleate dislocations.

concentrations. These are sufficient to initiate sources of dislocations in fairly soft metals and the new dislocations flow from these points in profusion. Slip occurs on two main planes, like ears, at forty-five degrees to the crack surfaces and so the worst of the stress concentration is relieved. This is roughly equivalent to rounding off the tip of the crack and thus, even if the Griffith energy balance (Chapter 5) is favourable to the propagation of the crack, the mechanism for implementing it may be frustrated for lack of a sufficient stress concentration.

The crack is thus not able to proceed under purely mechanical instigation and so the material is safe in perhaps ninety-nine per cent of practical cases. However, we have to remember in all materials work that there is no hard and fast dividing line between chemistry, physics and elasticity. Experts in their ivory towers like to make these distinctions but the bond between the atoms is not aware of them. A bond can be broken by chemical, physical or mechanical means or by any combination of these causes acting additively. A bond which is strained elastically is more susceptible to being dissolved chemically and physically. For this reason points of high stress are especially liable to attack by solvents or to corrosion by chemicals.

As we have said, there are many cases where the Griffith energy balance is usually still emphatically in favour of spreading a stationary crack. It is merely prevented from doing so by the reduction in the stress concentration at the crack tip which deprives it of a mechanism. Now although the stress concentration has been much reduced by the creation of numerous dislocations at the tip of the crack it has not been wholly eliminated and moreover the way of life of the local molecules has been considerably disturbed and all sorts of fine-scale strains remain. The bonds in this region are therefore more susceptible than elsewhere to attack by any aggressive solvent or chemical in which they happen to be immersed at the time. This is why metals which are tough in air or in other reasonably dry gases may crack when stressed for any length of time when wetted with sea-water or in chemical plants even though they may last for years submerged in the corrosive environment without a load. Some brasses are notorious traps for amateur constructors in this respect.

Ductility in crystals

The word ductility comes from the Latin *ducere* to lead, meaning, I suppose, that the material can be led by a stress to distort in a desired direction. In the engineering of metals it has two extremely useful consequences. It makes the metal tough so that cracks do not readily extend and, secondly, it may make the metal malleable (Latin *malleus*, a hammer), that is, able to be shaped, either hot or cold, by hammering, pressing or bending. Generally speaking, more capacity for flow is needed to obtain a useful degree of malleability than is needed to make the material reasonably tough. On the other hand a great deal of fabrication is carried out when the metal is hot when practically all crystals are much more ductile.

Ductility is exclusively a property of crystals for the reason that true dislocations can exist only as departures from the ordered crystalline state. Most solids are crystalline and dislocations exist in nearly all crystals. On the other hand, in the great majority of crystals, the dislocations are not sufficiently mobile at room temperature or are not mobile in the right way. Nearly all crystals contain quite large numbers of dislocations which arise from the nature of the mechanisms by which they grow but these dislocations are dispersed throughout the body of the materials in a very roughly uniform way. However, the stress concentration at the tip of a crack is a very intense and localized affair and there are generally not enough dislocations available naturally in the immediate vicinity to provide enough slip to relieve the situation, even if the dislocations are very mobile. It is therefore necessary that many new dislocations should be born on the spot, nucleated by the stress concentration itself. Furthermore, this must be able to happen very quickly if the material is not to be vulnerable to a sudden blow.

In real life, cracks are not two-dimensional diagrams on a sheet of paper but are flat, wedge-shaped holes trying to penetrate a solid three-dimensional material. To relieve the stress concentration adequately, there has to be slip on five planes in all.

The number of crystals which satisfy all these conditions at once is very few: out of all the thousands of crystalline substances which exist, perhaps only something like a dozen metals. With the rather dubious exception of silver chloride, no non-metallic crystal at present can be considered as truly and reliably ductile.

Although an enormous amount of research has been done on dislocations during the last thirty years and there now exists a really frightening volume of theory and information, it cannot be said that all the causes which determine the mobility of dislocations in different substances are fully understood. However, it may be worth looking at some of the more obvious reasons.

First of all, bonds vary a good deal in the ease with which they can be broken and reformed, and, of course, every time a dislocation jumps a step bonds have to be broken and remade. In this respect the most mobile bonds are those which exercise their attraction symmetrically in every direction, most notably the metallic bond and after this the ionic. The worst is presumably the covalent bond which is often highly directional and has an all-or-nothing character. Unfortunately the covalent bond is also the strongest and stiffest and the most generally desirable of the chemical bonds. Dislocations in covalent crystals are never usefully mobile at ordinary temperatures.

Again, the crystallographic structure of the crystal is important; that is, the geometrical pattern according to which the atoms or molecules in the crystal are stacked. If the unit cell or repeating pattern in the crystal is large, then a dislocation jump will generally be more difficult. Even if the unit cell is small but the packing is geometrically slightly more complicated, then the number of directions of easy slip may be unduly restricted. On the whole, crystals with cubic arrangements of atoms are more easily ductile than hexagonal ones. Furthermore the size of the crystals and the impurities they contain all have an important effect.

Although the vast majority of crystalline substances have no useful ductility at ordinary temperatures, those which do have tend to be altogether too ductile. Pure crystals of iron, silver, gold and so on, are too soft to be of much practical use and the art and science of metallurgy consists very largely in making such crystals harder and stronger without making them too brittle. This has to be done by controlling and restricting the movement of dislocations without stopping it too much.

Engineers are very apt to talk about and to specify 'elongation' as a measure of ductility. This is a rough practical test of the

amount by which the metal can flow before total fracture occurs
and has nothing at all to do with the elastic breaking strain of
the material which is usually somewhere below 1 per cent.
Elongation is measured, quite arbitrarily, by seeing how far two
marks on the stem of a test-piece, initially two inches apart, have
separated when the broken halves of the specimen are fitted
together after fracture. If the total distance between the marks is
then, say, three inches, the elongation is called 50 per cent, and so
on. As with most popular engineer's tests it is very difficult to
relate elongation in any consistent way either to the flow properties of the material or to the end usage. However, many
engineers have an almost religious belief in the value of the test
and if you tell them that wood and fibre-glass will give an elongation of nil but yet are very tough, they will merely reply that
that is why they do not use wood and fibre-glass. As with most
emotionally held beliefs this one presumably arises from fear, a
very reasonable fear of brittle failure.

In practice, with most metal alloys, an elongation of about 5
per cent or 10 per cent is usually sufficient to ensure a tough
material. The materials which are really popular, like mild steel,
may have elongations in the region of 50 or 60 per cent even
though the attainment of so much ductility implies the acceptance
of quite low tensile strengths. This is partly due to an attitude of
over-insurance against cracks but also to two other reasons. Many
structures are fabricated from sheets and bars and tubes of metal
and it is convenient and cheap to be able to bend these things to
shape in the cold. One can also use rather brutal methods to make
things fit. During the War I was told by one assembler of aircraft
that the only way in which he could get the wings of his Spitfires
to fit on to the fuselages was to bend the root fittings with a sledge
hammer. I never saw this done with my own eyes so that I cannot
guarantee that it is true but things of this sort certainly do happen
on occasion though perhaps not in the aircraft industry in peace-
time.

The second reason is that stresses can sometimes be relied
upon to readjust themselves within the structure. In a complex
structure it may be very difficult to calculate the loads in all the
various members with any accuracy, or perhaps one is just too

idle to do so. If the material yields and has a long 'plastic range', as it is called, then an overloaded member may stretch and be little the worse for it. Many engineers have a strong belief in the 'self-designing structure'.

These are very real benefits in an imperfect and commercial world and they go a long way towards explaining the immense success of mild steel, copper and soft aluminium. There are two drawbacks however. The ductility of even the softest metal is not inexhaustible and since there is generally no way of measuring how much of it has been used up in the manufacturing operations, one is left to guess how much of the initial ductility is left over to provide toughness in service. This is at the root of many complaints about failures in mass-produced goods. Annealing is a relatively expensive and troublesome operation, small components are costed and sold on tiny fractions of a penny and so the temptation to deform the metal in the cold as much and as often as possible may be irresistible. Perhaps it will not actually break until after the guarantee has expired.

The other drawback is that the requirement for maximum ductility necessarily implies a low tensile strength, because the metallurgist must arrange for dislocation movement to begin at a low stress. This has the consequence that structures are often much heavier than they need to be.

Dislocations, the edge and the screw

Dislocation theory is immensely complicated and, after all, perhaps mainly of interest to dislocation experts. We must however mention the two main varieties, the edge dislocation and the screw dislocation. The edge dislocation, which is that postulated by G. I. Taylor in 1934, is perhaps the easiest and simplest to think about. It has already been described in Chapter 4 (see Chapter 4, Figure 5). As we have said, it consists essentially of an extra sheet of atoms slipped into the crystal like a sheet of paper partially inserted between the pages of a book. Edge dislocations can be formed when the crystal grows, for instance at what are known as 'small angle boundaries'. That is, when two growing crystals meet at a fine angle so that they join up to form, effect-ively, one crystal, the join is marked by a line of edge dislocations, which may afterwards, of course, drift away.

Screw dislocations were postulated by Professor Frank, of Bristol, around 1948, not so much to explain the mechanical properties of crystals, as to explain their growth. For atoms or molecules to come out of solution or out of vapour and to settle down, more or less permanently, in a solid crystal, requires an energy change. Whether it happens or not depends upon what is called the supersaturation; roughly speaking, on how badly the molecules want to come out of solution or out of the vapour. One can cool solutions of sugar or salt, for instance, well below the temperature at which crystals should be deposited without any crystals appearing unless some surface is provided which is to their liking.

For an ordinary plane, flat surface the supersaturation which can be attained without actually depositing material can be calculated and is found to be quite high. What worried Frank was that, in practice, many crystals grew healthily at supersaturations very far below those calculated for deposition on a flat surface. In fact, if we had to depend upon deposition on a flat surface very many crystals could hardly be got to grow at all. It can be shown that if the surface possesses an irregularity such as a step, even if it is only one molecule high, deposition is much easier.

The re-entrant part of the step provides a welcoming and comparatively cosy home for wandering molecules which tend to settle down there, just as a bricklayer lays a brick in an existing re-entrant in a course of brickwork. As with a course of bricks, the result is not to abolish the re-entrant but to cause it to move along the top of the wall as more bricks are laid. This mechanism had been observed in operation by Bunn and Emmett around 1946. It will be remembered from Chapter 4 that this is the cause of the steps which mechanically weaken the surfaces of whiskers and other crystals (Chapter 4, Plate 10).

Frank's difficulty was that, admitting the existence of growth steps, what happened when a moving step came to the edge of the crystal? Presumably it would be extinguished, as the step in a course of masonry is extinguished when the bricklayer reaches the end of a wall. If so, then how could it be regenerated all over again for the next layer?

Frank's solution was brilliantly simple. Crystals are not built like a house out of level courses of bricks or molecules. The growth step was never extinguished at the edge because the crystal is built like a corkscrew staircase. Thus the crystal simply went on screwing itself into existence, using the same step indefinitely. Like G. I. Taylor's hypothesis of the edge dislocation, the screw dislocation is so intellectually satisfying that one feels it must be true and again, it turns out that it is. The screw dislocation was confirmed observationally by Forty and others not long after Frank conceived it (Figure 4).

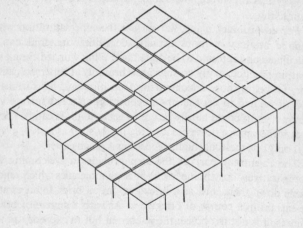

Figure 4. It is not very easy to visualize the arrangement of atoms in a screw dislocation. This diagram attempts to do so with rather moderate success.

The difficulty with corkscrew staircases is what happens in the middle. There is of course a hiatus or mismatch, in the form of a line of misfit up the middle of the screw and this is the actual dislocation. As with the edge dislocation the interatomic bonds are naturally highly distorted or strained at this point although in the normal way there is nothing one could call an actual hole. However it is not uncommon for whiskers to be hollow, like a tube. Possibly the reason is that these whiskers have grown with a screw dislocation of which the step is not one but several molecules high. If so, then the strain at the centre of the dislocation must be very large. The crystal may thus perhaps trade strain energy for surface energy, that is grow with a hole down the middle.

Like most successful hypotheses the screw dislocation has been rather overworked and almost every aspect of the growth of almost every kind of crystal has been attributed to it. It now seems that quite a number of crystals manage to grow without using the Frank mechanism but the fact remains that a very large number do use it and it is a very real and important phenomenon.

It is not necessary for dislocations to be wholly of the edge or wholly of the screw type. A single dislocation line can begin as an edge and end as a screw and vice versa. In between it partakes of the character of both and is spoken of as having a screw component and an edge component. However, the rules which govern the motion of the two types of dislocation are different and this is one of the reasons why the behaviour of real dislocations, which are usually lines curved in three dimensions, is very complicated.

The study of dislocations is now an elaborate and well-supported science in itself which has undoubtedly thrown a great deal of light upon the behaviour of solids in general and of metals in particular; to a considerable extent we now understand why metals behave as they do. On the other hand it cannot really be said that the knowledge of dislocations has led directly to any radical improvements in the mechanical properties of materials. As far as metals are concerned it does seem rather as if most of the possible important improvements have already been made by traditional and empirical methods and that the role of dislocation theory has been to explain the reasons for the improvements afterwards.

The energy of dislocations and the work of fracture of metals

Although it often requires only a small stress to move a dislocation through the body of a crystal – so that the strength of highly ductile metals is often very low – yet it will be realized that both the screw and the edge dislocation involve the existence of very high strains and distortions among the atoms in their immediate neighbourhood. Thus a considerable amount of strain energy is necessarily associated with every dislocation; this is known as 'the energy of the dislocation'. This energy, which does not differ very greatly between edge and screw dislocations, can be calculated quite easily and turns out to be about 10^{-9} Joules for each metre of dislocation length for most metals.

This amount of energy may not sound very much but it is to be remembered that, when a metal is extensively sheared, enormous numbers of new dislocations are created, in fact about 10^{16} per square metre. Thus the amount of energy stored in the dislocations in a cubic metre of highly strained metal is about 10^7 Joules. If we suppose the metal to be distorted to a depth of about one centimetre when it is fractured, which is roughly true, then this corresponds to a work of fracture of about 10^5 J/m²,* which is in fact the case for metals like mild steel. So everything adds up very nicely and that is where the work of fracture of a metal goes. Thus the dislocation mechanism not only acts as a crack-stopper, it also provides a very high and useful work of fracture.

The observation of dislocations

However plausible and intellectually satisfying a scientific hypothesis may be it remains an abstraction without subjective reality to most people unless one can actually touch or see it. Indirect or mathematical proof is not enough. Heat is a case in point. Everybody knows, from elementary physics, that the temperature of a substance is due to the motion of its molecules which are in perpetual but highly irregular movement. However, since one is also told that molecules are far too small to see and since the

---

*This value of energy is roughly equivalent to the area under the stress-strain curve.

sensations of heat and cold do not in any way resemble that of moving particles the idea of heat as molecular motion is usually not very real to us.

The botanist Brown discovered in 1827 that fine particles of pollen in certain flowers appeared under the microscope to be in perpetual dancing motion. The Brownian movement is most easily seen by making a suspension of fine scraps of solid in water. This can be done with ordinary Indian ink or with the water-colour gamboge. A drop of this can be put on a microscope slide, preferably covered with a cover-glass, and observed with a fairly high magnification in an ordinary optical microscope. The finer particles can be seen to be spinning and dancing in a most erratic jig for as long as one cares to watch them. What is happening is this. The particles themselves, which are perhaps a micron across, are a few thousand times larger than the molecules of the the liquid which surrounds them. These molecules are rushing hither and thither in a thoroughly random way. The particles of ink or gamboge are therefore being jostled in a rough and irregular manner. For the smaller particles these jostlings do not neces-sarily cancel out, as they do for large particles, and so the little particles are pushed or kicked around in a fashion which is visible in a quite ordinary microscope.

Once one has watched the Brownian movement one's appre-hension of the *nature* of heat will never be the same again. It is not that one can be said to have learnt anything in an objective scientific way but rather that one has come to terms with the kinetic theory of heat at a subjective level. It is the difference between having a sunset described and seeing one.

It is very much the same with dislocations. What began as an abstract hypothesis has become a very tangible phenomenon. What are the ways of observing a dislocation? Well, firstly one can etch it. As we have said mechanically strained bonds are more easily broken by chemical and physical means than unstrained ones. Thus one can prepare etches, usually acid brews, which will attack dislocations where they emerge on the surface of a crystal, in preference to the surrounding material. This produces a series of little pits which are usually easily seen in the optical micro-scope. This is a very common experimental technique and the expert can draw quite extensive conclusions from the series of pock-marks produced in this way on the surface of a crystal. One trick is to split a crystal in two. Any dislocations existing in the crystal before the experiment and which cross the cleavage surfaces will naturally be the same for both cleavage faces. One half of the crystal is selected as a control and etched immediately, thus showing up the pre-existent dislocations. The other half is distorted or otherwise experimented upon before it is etched. By comparing the etch-pits on the two surfaces one can see which dislocations were generated under the experimental conditions and also which have moved.

Etching is useful but it cannot be called a direct observation of the dislocation itself and so is perhaps not subjectively very satisfying. The next step in this direction was taken by Dr Hirsch in the Cavendish Laboratory, Cambridge. Very thin metal films are practically transparent in the electron microscope, but any distortion of the crystal lattice may show dark. The line of a dislocation thus shows as a dark line on a white background. So far so good, but one wants to see them move and it takes a stress to do this. It is not easy to apply a direct mechanical stress to films thin enough to be transparent to an electron beam. Hirsch however used the heat of the electron beam itself to expand the film and thus to stress it. This worked very well and he was able to film dislocations in motion. The impact of this film is considerable. The dislocations give an uncanny impression of scurrying mice.

In Hirsch's pictures, however, there was no question of seeing individual molecules or the three dimensional chequer-board of the crystal lattice. Hirsch's dislocations were simply moving black lines of strain on a white or grey ground. What we really want to see, I suppose, is a layer of molecules coming to an end in the crystal lattice. After all, if one saw a course of bricks doing this in a brick wall one would be highly surprised. However, in order to see a dislocation in a crystal lattice one must first see the crystal lattice.

Now in metals and in the majority of ordinary crystals the lattice spacing is about two Ångströms. At the time I am talking about, around 1955, the very best resolution of an electron microscope was about ten Ångströms so, in the ordinary way, there was no hope of resolving the layers in normal crystals.

Jim Menter at Hinxton Hall, near Cambridge, got over this difficulty by making and using thin crystals of a substance called platinum phthalocyanine. This is a flat, roughly square, organic molecule about twelve Ångströms across. In the middle of the square is a hole, and in the hole, in the case of platinum phthalocyanine, is an atom of platinum. (Copper phthalocyanine is a first cousin, the synthetic pigment Monastral Fast Blue, familiar in the blue paint on innumerable front doors.) In the crystal the flat molecules stack so that the lattice spacing is about twelve Ångströms and in the middle of each row of molecules is a line of heavy platinum atoms, standing out from the light atoms of the surrounding organic molecule. There are thus lines of platinum atoms in regular crystalline array, but spaced twelve Ångströms apart instead of the normal two. The organic part of the molecule may be regarded as a transparent padding or spacer keeping the dense, opaque, platinum atoms separated.

By adjusting the microscope to give the best resolution it proved possible to resolve the lattice of the phthalocyanine crystal. The result looked like rather woolly charcoal stripes on a lighter grey background very like the lines on a television screen. The first impression was that of the incredible regularity of this tiny scrap of crystal. On big enlargements the innumerable fuzzy black stripes extended perfectly straight apparently for ever and there were an enormous number of layers, millions upon millions of molecules, each one perfectly in its place.

Many pictures were needed before, after diligent searching, an edge dislocation was found. It looked exactly like the diagrams people had been drawing for twenty years. One dark fuzzy stripe came to an end and the others came together to close the gap (Plate 14). Jim Menter was able to send this photograph to Sir Geoffrey Taylor in time for his seventieth birthday.

To those of us working at Hinxton at the time, these pictures, coming wet from the darkroom, had a numinous quality not far short of a religious experience. A visiting Russian scientist looked at them for a long time and then he said 'You are looking up zee trousers of God.'

Jim Menter's visual revelation of the crystal layers and their dislocations by means of the electron microscope is satisfying and has become famous. There is however another approach with a strong subjective effect. As we said in Chapter 4, David Marsh devised a most sensitive tensile testing machine for whiskers and other fine fibres. This machine can detect extensions of as little as four or five Ångströms, which is about the resolution of a modern electron microscope. The shear slip movement caused by a single dislocation is about one Ångström and therefore cannot be measured by the machine. A dislocation source however releases sufficient dislocations to provide from one hundred to about five hundred Ångströms movement and thus can easily be detected by the Marsh machine.

Now if we take any ductile material, such as a soft metal, and pull an engineering-sized test-piece we shall get a load-extension diagram such as Figure 5, a smooth curve of a type very familiar to engineers and metallurgists.

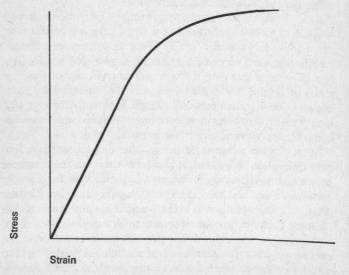

Figure 5. Normal macroscopic load-extension curve for a ductile substance.

If we now take a really thin but ductile test piece, such as a large whisker, and test it in the Marsh machine, we get a load extension diagram which is something quite different. A typical example is that in Figure 6. Here we have an elastic extension

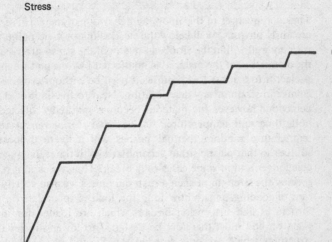

Figure 6. Similar material tested on a fine scale in the Marsh machine. Plastic extension is broken down into steps, each of which corresponds to the operation of a dislocation source.

interrupted by the sudden operation of sources. These sources operate quite erratically and the slip they produce is virtually instantaneous. The load extension diagram therefore has the form of a series of steps. What is happening is that, at each level of stress there are dislocation sources almost ready to give birth. What triggers them into doing so is a random thermal shove, just like the shove or jostle which pushes the particles around in Brownian movement. In a large specimen the same thing is happening but in so many places and so frequently that the gross effect is that of a smooth curve.

With a small specimen in the Marsh machine the complete randomness and suddenness of the movements yet again impresses upon one the reality of dislocations.

Creep and temperature resistance

The consequences of this upon the behaviour of metals in service are fairly obvious. Well below the elastic limit or yield point, that is to say well within the Hookean region of the stress-strain curve, the elongation of the material is unaffected by time and we might subject it to a stress for centuries if need be without causing any change in strain or any deterioration. Near to the limit of elastic behaviour however the material becomes markedly affected by both time and temperature. As we have seen, even at room temperature random thermal pushes will activate dislocation sources so that plastic strain accumulates with time: the material gets longer and in some cases may break. Thus we cannot really specifiy the strength of such a material unless we also specify the rate of loading or say how long the load is to be left on for. Structures like suspension bridges which are loaded for many years on end must therefore be designed to lower stresses than structures which are only stressed rapidly and occasionally. In practical metals there is often *some* creep even at quite low stresses and this has to be watched where dimensional accuracy is important.

As we might imagine, the stress at which creep becomes serious is very dependent upon temperature and this is a factor which frequently governs the design of machinery and especially of heat engines such as gas turbines. On the whole the hotter the hot parts of an engine can be run the more efficient the engine is likely to be, especially in the matter of fuel economy. Since iron melts at over 1,500° C. and other metals at even higher temperatures it might be thought that there would be no great difficulty in running engines at temperatures of, say, 1,200° C. which are well below the melting point. This is very far from being the case.

It is true that iron does not melt below 1,500° C. but then the concept of melting implies that the metal flows under its own weight which is usually a negligible stress. As soon as we introduce a mechanical stress, however, even a small one, flow and eventually fracture occur at far lower temperatures. Even for comparatively rapid loading, that is to say loading as rapid as can conveniently be applied in a testing machine, the strength is drastically reduced. Furthermore, when components are subjected to prolonged stress in one direction, such as the centrifugal stress in turbine blades, we have to be very careful indeed about the creep.

For 'short-term' loading the strength of a metal varies with temperature much like Figure 7, in other words the material dies

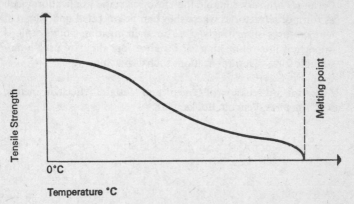

Figure 7. Most metals get weaker when they are heated and the weakening effect begins to operate surprisingly early. Thus although the melting point of iron is 1535°C. many steels cannot be used above about 300°C.

with a whimper not a bang. As a very rough working rule, which is true for most metals, one cannot work a metal at a temperature which is higher than half its melting point in degrees Kelvin. (Degrees Kelvin = degrees Centigrade+273. See Appendix.)

The way to get the operating temperature of an alloy up is, of course, to impede dislocation movement by one means or another. The difficulty is that most of the additions which one might use for this purpose tend themselves to become mobile. (Naturally one must avoid oxidation of the surface of a high temperature alloy but the additives introduced for this purpose may conflict with those intended to prevent creep). The subject is an extraordinarily difficult one and metallurgists have probably done well to achieve working temperatures as high as 950–1,050° C., which can just about be done with the recent very special turbine blade alloys. The normal run of engineering steels have to be operated far below this temperature.

It is true that dislocations are largely immobile in many ceramics at temperatures up to about 1,500° C. but then these materials are usually dangerously brittle at lower temperatures. Ceramics are very suitable therefore for static applications such as furnace refractories where they can be operated under load at temperatures surprisingly close to their melting points (this is important in steel-making for instance) but they are usually unsuitable for moving applications such as machinery.

Note: for a discussion of 'creep' and 'fatigue' effects in metals, see *Structures*, Penguin Books, 1978.

Chapter 10   Iron and steel

Hephaistos among the Satanic Mills

> '*I will send them the locomotive to be the Great
> Missionary among them.*'
> George Stephenson (1781–1848).

Simply because wood and masonry have always been the commonest materials of construction and metals are relatively recent intruders in technology and engineering we are peculiarly conscious of the role of metals as innovators although, as we have seen, they are still in a minority on a tonnage basis. However, metals, especially iron, were remarkably well suited for making the sort of machines which were both the curse and the glory of the Industrial Revolution. It was the development of means of making and working iron cheaply and in quantity which made the spread of mechanization possible. Steel, as we know it, only came in as a cheap material in the second half of the nineteenth century when the main impetus of the revolution was over.

Important as iron was, it was only used to the smallest practicable extent in most of the early machines, even in steam engines. In Fulton's first steamship on the Hudson River even the boiler, incredibly, was of wood; the actual heating of the water took place in a separate arrangement of iron pipes. Even the Americans thought that this was going rather far, although the river steamers which carried the bulk of the American inland traffic until after the period of the Civil War used wood in their machinery, as well as in their hulls, to a degree which a modern engineer finds hard to credit.

Broadly speaking there are two problems with all metals – extraction metallurgy, the separation of the metal from its ores; and physical metallurgy, which is how to get the metal into the most useful condition of hardness, strength and toughness. As we have seen, a pure metal is usually very soft so that physical metallurgy consists, notionally, of impeding the dislocations to just such an extent that the metal is strengthened without being embrittled. Technologically however the processes of extraction metallurgy frequently do not yield the virgin metal in a pure condition so that both extraction and the subsequent processing have to be considered together.

In their chemical and technical details the treatment of the different engineering metals varies widely but in every case the physical objective remains the same – the control of dislocation movement by adjusting the size and structure of the metal crystals (dislocations can cross grain boundaries but they do so with some reluctance) and by adding other substances and 'impurities'. Complex alloy systems may ensue in which it is hoped that the diversity of crystal structures will impede the dislocations by just the desired amount. Very small particles, even single impurity atoms, will 'pin' the forward movement of a dislocation line wherever it meets them. The stress required to bow out and eventually to detach the dislocation line between the pinning points is calculable and depends upon the separation of the points and therefore this is a useful and powerful means of control. Almost anything which is added to a metal will affect its mechanical properties for good or for bad – sometimes for both since one has to be careful that an additive which is beneficial when dispersed does not end up concentrated around the grain boundaries where it may have a serious weakening effect as we saw in Chapter 4.

When we consider the number of ductile metals and also the number of hardening mechanisms which are available the permutations and combinations of physical metallurgy are very numerous and the subject is a most complex one. To the non-specialist, however, who is chiefly interested in principles and in the end results, the consequences of these processes are very much to a pattern.

Many of the engineering metals are listed in Table 1 of Chapter 8 where it will be seen that – although there is a very considerable range of specific gravities, from about 10·5 for molybdenum down to 1·7 for magnesium – for all the metals in that table the Young's modulus divided by the density of the metal comes out to an almost constant figure, namely $3·9 \times 10^6$ p.s.i. (25,000 MN/m²); there are a few common metals not in the table, such as copper and brass, which have specific moduli which are a little lower than this. There is no ductile metal with a higher stiffness for its weight and thus nearly all the useful metals give one in stiffness exactly what one pays for in terms of weight, neither more nor less.

As has been said, all these metals are very soft and weak when in the pure state and it is the business of metallurgy to raise the strength and hardness without rendering them unduly brittle. Seen as *strain* the achievements of metallurgy are remarkably constant. Usually the maximum *elastic* strain which can be imparted without unduly embrittling the metal is around 1·0 per cent, give or take a little. However engineers generally regard metals in this state as having too little ductility for most purposes and for the majority of uses they are satisfied with a maximum elastic strain between 0·25 per cent and 0·5 per cent, when the elongation is likely to be around 50 per cent to 60 per cent.

Very, very roughly therefore all the metals may be taken as constituting a family with similar specific stiffness, specific strengths and elongations. This generalization is only approximately true and it is not quite fair to metallurgists who do, as a result of their labours, continue to attain rather better combinations of specific strength and toughness (specific stiffness they can do nothing about) although the limits are fairly narrow. What metallurgists have been more successful at is making metals retain their room-temperature properties up to elevated temperatures. For many uses of course this is more important than getting more strength at room temperature.

It is not necessary to describe here the special metallurgies of all the various engineering metals. There are innumerable books on the subject including an excellent general review of technical metallurgical processes in *Metals in the Service of Man* in the Pelican series. However, the great social and technical importance of iron and steel renders some account of these materials necessary. In attempting it I am only too well aware of the size and difficulty of the subject. Perhaps before doing so I should have sacrificed to Hephaistos, smith and purveyor of weapons to the Olympians, the only technologist to have reached the rank of a major God.

Iron

The elementary facts about iron and steel tend to be obscured by the jargon of the trade. First of all there is a confusion about what is meant by 'iron' and 'steel'. Naturally in both cases the predominant chemical element is iron. 'Iron' by itself usually means iron in the relatively pure chemical form. 'Cast iron' on the other hand means iron containing nearly as much carbon as it will hold, perhaps about four per cent. 'Wrought iron' is different again and is usually a special sort of fairly pure iron containing glassy inclusions. 'Steel' usually means iron with a little carbon in it, generally less than 1·0 per cent. The implication is that in steel the content of carbon or other alloying element is under fairly close control. Since the mechanical behaviour of iron is profoundly influenced by very small amounts of carbon and other additives the control which is exercised over steel is important. 'Alloy steel' is usually a generic term for steels which are alloyed with elements other than carbon. Alloy steels, incidentally, are generally a good deal more expensive than carbon steels.

The strength of iron and steel is governed by the extraordinary sensitivity of dislocation movement within the iron crystal to quite small traces of carbon. Of course, dislocation movement was not understood until recently and even the comparatively simple chemistry of the extraction metallurgy of iron was not appreciated until fairly late in the Industrial Revolution. However, the practical metallurgy of iron was worked out without benefit of these ideas and it remains today largely a traditional process. Just as, in the textile trade, spinning and weaving go back to prehistoric times and all that the most advanced factories have done is to mechanize and rationalize the simple hand process, so the steel trade operates techniques which are sophisticated versions of immemorial practices. On the whole therefore the iron and steel making processes are best understood in their historical contexts.

Apart from his scientific ignorance, the biggest difficulty that the primitive metallurgist had to encounter was that of getting a high enough furnace temperature. In a modern steelworks the availability of high and controlled temperatures enables short cuts to be taken which by-pass a number of ancient procedures. Naturally everything is on a much bigger scale and a modern furnace may produce a thousand tons a day where the medieval smith would have been content with a hundredweight.

Unlike bronze which can be melted at about 900° to 1,000° C. – which is just within the reach of an ordinary wood fire – pure iron melts at 1,535° C. which for long centuries was out of the range of human technology. However, the addition of quite small proportions of carbon lower the melting point of iron considerably and, of course, carbon is readily available when a carbon fuel, such as charcoal, is used to heat iron ore. The lowest melting temperature attainable in this way is about 1,150° C. which occurs when 4 per cent to 4½ per cent of carbon has diffused or seeped into the metal.* If a mixture of iron with excess carbon is heated to 1,150° C. therefore, it will melt. Such a temperature was not very easy for primitive people but it could just about be attained by a blown charcoal fire.

Iron ores are usually oxides of iron, typically haematite, Fe_2O_3, which is so called from its blood-red colour. The iron oxides, incidentally, are used as pigments in painting; Venetian red is largely Fe_2O_3. The first thing to do is clearly to get rid of the oxygen. When the ore is heated with charcoal or coke this occurs almost automatically:

$$3Fe_2O_3 + 11C \rightarrow 2Fe_3C + 9CO$$

The oxygen goes off with some of the carbon as carbon monoxide gas, leaving iron carbide or cementite, a compound containing 6·7 per cent of carbon. However in practice we also get the reaction:

$$Fe_2O_3 + 3C \rightarrow 2Fe + 3CO$$

So some pure iron is produced as well as cementite and we

*The amounts of carbon present in iron and steel appear surprisingly small considering the effect which they have upon the material. It is to be remembered however that the proportion of carbon in iron is always quoted as a percentage by *weight* and that the carbon atom is much lighter than the iron atom – roughly a fifth of the weight. The percentage of carbon by volume, or by numbers of atoms, is thus considerably larger and may reach about twenty per cent.

usually end up with a mixture of iron and iron carbide containing, as a whole, 4 per cent or so of carbon. Iron and iron carbide are mutually soluble and it is this solution which forms the liquid whose low melting temperature was the key to the extraction of iron in primitive furnaces – it is also what the modern blast-furnace produces.*

Iron ores do not consist solely of iron oxide but contain various mineral impurities, mostly oxides of other metals. These have usually, by themselves, also a high melting temperature and if the ore were heated solely in contact with a carbon fuel there would be a danger that it would not melt properly. To ensure that it does so a 'flux' is added, usually lime or limestone. This lime performs exactly the same function as it would in glass-making, that is to say it reduces the melting temperature of the non-ferrous oxides by forming with them a glass with a lowish melting point. This is called slag. In commercial practice it is a dirty brown or grey substance which is nowadays sometimes made into fibres and sold for insulating houses.

What we are apt to get at the bottom of the furnace is therefore a mixture of iron, iron carbide and slag. In the most primitive processes this mixture could hardly be melted and was extracted as a cake or 'bloom' containing bits of charcoal and other impurities. Apart from the question of these impurities, iron carbide is a brittle substance unsuitable for making tools and weapons. This is because, while crystals of nearly pure iron are held together by metallic bonds which favour the passage of dislocations, crystals of iron carbide are held together by bonds which are partly covalent in character so that dislocations do not

*This description applies to normal extraction processes operated at or above 1150°C. The earliest iron-making seems to have been carried on at considerably lower temperatures, perhaps as low as 700°C. Under these conditions no melting takes place but the *second* of the two reduction reactions takes place, rather slowly, in the solid state. The initial product of this process is a bloom consisting of iron which, chemically speaking, is fairly pure but which is of course full of bits of charcoal and other forms of dirt. This bloom is also beaten or wrought with hammers, in this case partly to get rid of the larger impurities but also to disperse carbon through the metal. This leads eventually to some sort of mixture of iron and iron carbide which can be hardened by quenching.

become appreciably mobile until a temperature of 250° C. or so is reached. The metal in the fresh bloom is thus malleable when hot but brittle when cold.

The early smiths therefore took the crude iron from the extraction furnace and, after reheating to some temperature perhaps about 800° or 900° C., they hammered it. Originally the iron was 'wrought' or hammered by hand, with immense labour, but latterly this was done by water power, supplied by 'hammer ponds'. The hammering had two effects. It squeezed out mechanically most of the impurities and some of the slag and it also reduced the carbon content of the iron. This occurred in the following way. Iron heated in air to a moderate temperature forms an oxide scale, the commonest form of which is FeO. Iron which is heated and beaten out flat thus becomes covered with oxide and, when, after it has been beaten into an elongated form, the smith doubles it over like pastry and beats it out again, the film of oxide is included between the layers of hot metal with which it is hammered into intimate contact, so that the simple reaction occurs:

$$Fe_3C + FeO \rightarrow 4Fe + CO$$

In high-grade work the beating out and folding over were repeated, sometimes for thousands of times. This is why swords show a delicate wavy pattern, each line corresponding to a thin layer of metal and to a beating operation. If the job was properly done almost the whole of the carbon was removed, leaving iron which was nearly pure except for a little silicon which was on the whole beneficial. This 'wrought iron' contained however streaks and strings of slag which again were to some extent beneficial. This was because the purified iron was now generally too soft and the glassy filaments limited flow to some extent. Furthermore, the rust resistance of wrought iron is generally excellent. This is partly due to the purity of the iron itself but many people hold that the initial film of rust is anchored to the surface of wrought iron by means of the slag inclusions so that it remains to form a protective skin instead of dropping off to make way for fresh corrosion as it generally does with steel.

Wrought iron direct from the anvil was too soft for weapons

and cutting tools and therefore for these purposes it had to be hardened by putting back a certain amount of carbon, at any rate into the surface. This was done by a process almost identical with that which is still widely used, that is to say 'case hardening'. The sword or other weapon was packed inside a mass which consisted essentially of carbon but which often also contained a number of secret ingredients of dubious efficiency. It was heated in this environment for a period so that carbon diffused into the surface to a depth of perhaps half a millimetre to a millimetre.

This surface carbon hardened the metal considerably but, to get the best effect, the 'steel' might be quenched by cooling it suddenly in a liquid. The exact mechanism of this quenching is complicated. Briefly, the hot steel consists of 'austenite' which is a solution of cementite in iron which is unstable at room temperature. The way in which the austenite parts with the excess carbon when it is cooled depends on the details of the cooling process.

If cooling is comparatively slow the result will generally be 'pearlite', so called from its iridescent appearance in the microscope caused by its banded structure. Pearlite consists of alternate layers or laminae of pure iron (ferrite) and iron carbide (cementite). In this form and because of its regular structure, the steel is tough and fairly strong but not particularly hard. However, if the austenite is cooled very quickly the result will be mainly 'martensite'. Martensite is yet another variant of the iron-carbon crystal which has the carbon atoms squeezed in in such a way that dislocation movement is impossible and so the crystal is extremely hard. On cooling, austenite transforms to martensite at a very high speed indeed, about three thousand miles an hour, and so to get as much martensite as possible, it is necessary to quench as quickly as one can.

Quenching can be done in water and it usually is but, historically, there seems to have been a preference for using urine and other biological liquids.* It turns out that this practice really was

---

* 'Another sort of tempering of iron is also made in this manner, by which glass is cut and also the softer stones. Take a three year old black goat, and tie him up for three days within doors without food; on the fourth day give him fern to eat and nothing else. When he shall have eaten this

beneficial for two reasons. The first reason is that the cooling was rather quicker. When water is poured on hot metal a film of steam is formed so that the liquid water does not actually touch the metal and the heat transfer is consequently bad. This is easily demonstrated by dropping water on the hot plate of an electric cooker. With urine, however, crystals are formed on the surface of the metal as the water evaporates and these tend to bridge the steam gap and improve the heat flow. Furthermore, urine contains urea and ammonia both of which are nitrogen compounds. There was therefore some degree of nitriding of the surface, that is nitrogen diffused into the iron. This formed hard needle crystals of iron nitride, Fe_2N, and, also, individual nitrogen atoms insinuated themselves into the iron lattice as what are known as interstitials. Interstitial atoms pin dislocations. As a matter of fact the degree of nitriding effected during quenching is very small. In modern commercial practice periods of two or three days in urea or ammonia are needed, for this reason it is a rather expensive treatment, used where only the best will do.*

It will be noticed that the preparation of iron and steel consists of a series of approximations, each process going too far and being in turn corrected. Thus we first make cast-iron or pig iron which is too hard and contains too much carbon, then we generally remove all the carbon and find that the iron is too soft so that we have to put some carbon back again. This iron or steel, as has been explained, is generally 'quenched' by cooling suddenly in a liquid when we want a hard tool or weapon.

---

for two days, on the night following enclose him in a cask perforated at the bottom, under which holes place another sound vessel in which thou wilt collect his urine. Having in this manner for two or three nights sufficiently collected this, turn out the buck, and temper thine instruments in this urine. Iron instruments are also tempered in the urine of a young red-headed boy harder than in simple water.'

Theophilus Presbyter (eleventh century), *Scheme of various arts*, translated by Hendrich, 1847.

Note that 'tempering' is often popularly used when 'quenching' is really meant. Quenching is a hardening process, tempering tempers or softens extreme hardness.

* Of course the iron has to be hot for the nitrogen to enter the metal. Dogs do not harden lamp-posts.

Quenched carbon steel (or carburized iron) is often too brittle and so yet another and this time final process is often needed, that is tempering.

For tempering, the quenched metal is reheated to some temperature between about 220° C. and 450° C. and allowed to cool naturally. This softens the steel to some extent by transforming some of the martensite into a softer, more ductile structure. The higher the tempering temperature the greater the effect. Traditionally the temperature for tempering was judged by the colour of the oxide film found on the surface of the metal which varies from yellow through brown to purple and blue. It will be obvious that simple carbon steels of this kind cannot be operated at high temperatures without spoiling their properties.

Cast iron and pig-iron

As we have said, the earlier extraction furnaces could barely melt the iron which they made and this iron was usually removed from the bottom of the furnace as a rather messy cake or bloom. By the middle of the fifth century B.C., that is about the time of Pericles, Greek furnaces were able to melt the iron properly and so run it out into moulds as 'cast-iron'. Although cast iron was available through the classical period, the uses for such a brittle material were limited and it does not seem to have become economically important. Most classical iron is wrought iron.

With the dark ages the temperature of furnaces fell and cast iron does not seem to have been made again in Western Europe at least until the thirteenth century. It really found its métier however after the invention of gunpowder. The earliest cannon (the word comes from *kávva*, a reed or bamboo, and is connected with the ecclesiastical sort) were made from wrought iron staves bound with iron hoops, like a barrel. However, as the capacity of furnaces and the skill of ironfounders increased cannon began to be cast. The early cast guns burst nearly as often as the built-up wrought iron ones but they must have been much cheaper.* In

---

*As a matter of fact cast-iron guns never did become very reliable. *Victory's* opponent at Tragalgar, the *Redoubtable*, burst two guns before she surrendered. The guns were little better in the Crimean War. See page 249.

its traditional form cast iron is not only very brittle but it also contains little layers or veins of carbon in the form of graphite which act as built-in cracks. As a result, cast iron was both weak and unreliable in tension. It was therefore really a very unsuitable material for a pressure vessel like a gun barrel. However until about 1860 it was the only economic one available, brass or bronze being usually too expensive. In consequence, guns were extremely heavy; one throwing a thirty-two pound shot, such as the main armament in H.M.S. *Victory*, weighed between three and four tons. The weight of the guns therefore formed about ten per cent of the displacement of a warship. Because of her age, *Victory* can no longer bear the weight of her own guns and they have been replaced with wooden replicas.

At one time it was usual to cast iron directly from the blast furnace but this is not very often done today. This is partly because blast furnaces have become much bigger and it would not be economic to cast iron from them in penny packets, and partly because iron cast straight from the blast furnace is hard, brittle and weak. Nowadays, it is usual to cast the whole of the production of the blast furnaces as pig-iron. Part of this goes on to be turned into steel and part is remelted and its composition modified so as to produce cast iron with more acceptable properties. By taking thought, it is now possible to produce cast irons which are reasonably tough and have quite good tensile strengths. As it is generally cheaper to make a complicated shape by casting it in iron than by forging it in steel there is still a considerable incentive to improve cast iron which is nearly always used for the cylinder blocks of car engines, for instance, because of their elaborate shape.

In England iron was originally extracted with charcoal in the forests of the Weald and later in Shropshire, but during the first half of the eighteenth century the problems of using coke in place of the increasingly scarce charcoal were gradually solved. The change to coke was fairly complete in England by about 1780 although it did not take place on the Continent until much later. This not only enabled more iron to be extracted more cheaply but it was one of the reasons why the iron trade gradually migrated to what we now call the 'Industrial North'.

Thus, by the late eighteenth century large castings, up to about 70 feet long, were available in England quite readily and cheaply and they could generally be transported by water. These castings however were very weak in tension by modern standards and so they had to be used in applications where the stresses were predominantly compressive. An obvious use was bridge-building. Here the iron could be used in compression to make arches, much like stone arches. A stone arch is built up from wedge-shaped stones called 'voussoirs' each of which has to be cut laboriously by hand. The early iron bridges were made from cast voussoirs in the form of open or lattice frameworks which fitted together like a stone bridge.

The famous Iron Bridge, erected over the Severn at Coalbrookdale in 1779, was the first large iron structure and is more or less of this type. It has a span of just over 100 feet, a total length of 196 feet and rises 50 feet. It contains $378\frac{1}{2}$ tons of iron and was built in three months. It cost £6,000 to build and, even at the values of 1779, this was much cheaper than such a bridge could be built in any other material.

Although the Iron Bridge was successful, it gave trouble from its very virtue. An arch bridge, like an arch in a cathedral, thrusts outwards at the base, and whereas in a church this thrust may be opposed by buttresses, in a bridge the outward forces are balanced by the opposite push of the masonry and earthwork approaches. We are not apt to think of cast iron as a light-weight material, but compared with traditional building materials it is so. As a result, the arch at Coalbrookdale showed, perhaps for the first time in history, the opposite fault to that of stone arches and domes. It was too light to oppose the inward thrust of the approaches whose weight, tending all the time to slide into the river, forced the iron arch inwards and upwards. For this reason the conventional approaches had to be replaced by cast-iron subsidiary arches. This must have been an early case of that difficulty of putting new wine into old bottles which constantly troubles materials engineers.

Puddled wrought iron

After coke began to be used in mechanically blown blast furnaces crude cast iron became relatively cheap and plentiful though its uses in this form were limited by its brittleness and low tensile strength. For most purposes the stronger and tougher wrought iron was needed and as long as this had to be hammered out laboriously, wrought iron remained, even with water-driven hammers, a scarce and expensive substance. If one had to pick upon any one material as being the key to the Industrial Revolution, then that material must surely be puddled wrought iron. Steel in its various forms did not appear in quantity until much later and its social and economic implications were less important.

Puddling, at least in any practical form, seems to be due to Henry Cort (1740–1800) who patented the process in 1784. Cort invented a coke-fired furnace in which the chimney ran for a short distance horizontally before turning upwards in the usual way. On the bottom surface of this horizontal section was a basin shaped hollow in which could be melted a pool or puddle of pig-iron. This puddle of molten iron could be stirred through ports in the side of the furnace by long iron tools called 'rabbles', shaped rather like a hoe.

When the pig-iron was molten the puddler stirred iron oxide into it with his rabble. This oxide, often the scale from the rolling mills, when well stirred reacted with the carbon in the pig-iron in much the same way as it did in the hammering process, removing most of the carbon as carbon monoxide. The evolution of gas agitated the bath into a 'boil' which drove most of the slag out of the furnace. As the carbon was removed, so the melting point of the iron increased, and, as the furnace temperature was around 1,400° C., the iron began to 'come to nature' or to grow pasty. It was then rolled up into a ball weighing about a hundred-weight and removed from the furnace. Although puddling was very hard work a puddler could puddle about a ton of iron a day which was perhaps a ten- or twenty-fold increase on the hammering process. Puddling was a very skilled trade and for many years after the Napoleonic wars English puddlers earned good money by travelling around on the Continent giving instruction in puddling.

When puddled, the hot pasty iron was usually passed through rollers which, after many passes, squeezed it into plates or rods. In the process the hot surface oxidized and the resulting scale fell off as the iron cooled and was fed back into the puddling furnace. As will be seen, the whole process was chemically equivalent to beating out the iron in the old way but was considerably more productive. Nowadays puddling is almost extinct because, even with mechanization, the output of a puddling furnace can only be raised to about a hundred tons a day whereas, by blowing air through the iron in a Bessemer converter, such as is used for making steel, an output of about 800 tons per day is possible. In any case the market for wrought iron nowadays is limited since steel is both cheaper and stronger.

Many of the problems of engineering are really concerned with how much strength and how much toughness one can get for how much money. The whole of the Industrial Revolution has to be seen and judged against a background of the gradually falling price of wrought iron and mild steel. This is very clearly illustrated in the history of the railways.

Railways began as colliery tramways in which wooden rails were laid to ease the passage of horse-drawn waggons. Towards the end of the eighteenth century many of these wooden rails were replaced by cast iron ones which lasted longer and showed less rolling friction than wood, so that a horse could pull four or five loaded waggons on the level. For the transport of minerals this was considered quite satisfactory and probably no further modifications would have been made had it not been for the sharp rise in the cost of horse fodder at the time of the Naploeonic wars. This turned the minds of colliery owners to the possibility of using the coal which they produced themselves at 3/9d. per ton as a source of tractive power. The pit owners, of course, already used steam engines extensively for both pumping and winding but these stationary low-pressure (3 p.s.i.) engines were far too large and heavy in relation to their power output ever to become mobile.

The inventor of the high-pressure locomotive, and therefore

the true father of railways, was Richard Trevithick (1771–1833), a genius who died in poverty, unlike the Stephensons who both lived and died in the odour of great prosperity. Trevithick produced a high-pressure (50 p.s.i.) locomotive in 1804 and another in 1805 (Plate 15). Both were successful as locomotives. It was the track which let them down.

In spite of the relative costs of hay and coal, locomotives were expensive both to build and to run. The annual running costs of an engine, including capital charges and so on, were estimated at a little under £400, which of course was far more than that of a single horse, even if the cost per horse power was less for the engine. To be economic the locomotive must therefore either draw a greater load than the horse or else draw the same load faster. Since horses were already working along the line it was impracticable to increase the speeds very much and hence the engine must be made to pull more trucks.

As we know there is no particular difficulty about getting enough adhesion between smooth metal wheels and rails to draw any desired load, always provided that the weight on the driving wheels is sufficient to prevent them from slipping and this was where the real impediment lay. The existing cast-iron rails were just sufficiently strong to support reliably the three-ton trucks in use but an engine weighing only three tons itself could not be made to draw thirty or so trucks of the same weight without slipping. If the engine were made heavier then it broke the cast-iron rails so frequently as to be uneconomic. The failing of Trevithick's engines was that they broke rails in a wholesale manner and so they had to be converted to stationary uses.

After this the story of the development of the early locomotive is that of a struggle for adhesion without breaking the rails. Part of the trouble was that the early engines were not sprung because no strong enough steel springs could be made. In consequence the load on the rails was multiplied at every jolt. Engines were built with eight driving wheels in order to spread the load (Plate 16), but one of the most popular solutions was to cast the rails with teeth which meshed with cogs on the engine like a modern mountain railway (Plate 17). These devices were troublesome and never worked really well.

George Stephenson met the spring difficulty by providing his engines with 'steam springs', that is by supporting the axles on pistons floating in cylinders filled with live steam, a suspension exactly similar in principle to that recently introduced for motor cars. However, because of the problem of sealing the pistons properly Stephenson abandoned steam suspension as soon as steel springs became available.*

In 1821 John Birkinshaw of Bedlington near Morpeth patented a method of rolling puddled wrought-iron rails of I section and one of his early customers was George Stephenson who was thoroughly worried about the track for the Stockton and Darlington line at the time. Birkinshaw quoted £15 per ton for his wrought-iron rails and, although this was more than twice the cost (£6–15) of cast iron, the effective cost per mile turned out to be the same because, the wrought-iron rail being stronger, a lighter rail could be used. The length of Birkinshaw's rails was 15 feet which, it will be remembered, was the length of the forged bars made at Akragas in 470 B.C.

The American approach to the same problem a few years later was different. In many cases they seemed to have reverted to a system which was used in Scottish colliery lines about 1785. This involved laying a flat strip of wrought iron on top of a substantial wooden rail. This was used in America for some years and worked fairly well. The iron strip was however merely nailed or spiked to the wood beneath and from time to time the butt joints worked loose. When this happened the end of the strip might curl upwards under the weight of a train wheel and when this happened it could penetrate the floor of a carriage above it, sometimes with fatal results to the passengers.

Such tracks were replaced by the usual wrought-iron rails but for a great many years the American practice was to use rails of much lighter section than in Europe, supporting them by very closely spaced wooden sleepers. This habit of course reflected the cheapness of timber and the high cost of iron in America.

Although cheap steel became available about 1860 it took nearly thirty years to oust puddled wrought iron. In 1883 some

---

*One of the revolutionary features of the Rocket (1829) was the introduction of piston rings in the driving cylinders to replace oakum packing.

seventy per cent of the pig-iron produced in Great Britain was puddled, and it is doubtful if more than ten per cent was turned into steel. By the 1890s these proportions were roughly reversed. The real reason why puddled wrought iron hung on for so long was that, although weaker, and sometimes more expensive than steel, it was considered, with some justice, to be more reliable.

The first sensational success for the new Bessemer steel occurred when fast paddle steamers, such as the famous *Banshee*, were built in the early 1860s to run the Northern blockade into Southern ports during the American civil war. This they did with almost contemptuous ease, having a speed of twenty to twenty-two knots compared with the fifteen or so of the fastest ships of the Northern navy. Some of these vessels ended their days, comparatively recently, as passenger steamers on the Clyde. Although the saving of weight, by the use of steel, was very great, accidents occurred quite frequently and the British Admiralty would not build hulls in steel until about 1880. The use of steel for really large and important structures could not be said to have become established until the Forth railway bridge was built of open-hearth steel in 1889.

Steelmaking

The making of steel, especially under modern conditions, is an extraordinarily complicated business and only quite a brief outline can be given here. What we now call mild steels and carbon steels consist of iron having between 0·1 per cent and 0·8 per cent carbon content, little or no slag inclusions and with or without controlled small amounts of other elements such as silicon and manganese. What the traditional producer of wrought iron made however was iron almost free from carbon but containing extensive slag inclusions and also small but uncontrolled amounts of other elements. As we have said the biggest difficulties faced by the early ironmasters arose from the fact that, as they removed the excess carbon from the pig-iron, they also raised the melting point from about 1,150° C. to around 1,500° C., a temperature beyond the capacity of their furnaces. Thus slag could not be removed by melting and carbon could only be put back, to give the necessary strength and hardness, by carburizing the hot but still solid surface of the iron by blacksmith's methods.

However, throughout the eighteenth century the temperature of furnaces was slowly rising and about 1740 Benjamin Hunt found that he could melt wrought iron with a little added carbon in small batches of up to about 90 lb. in covered clay crucibles in a furnace. When this was done, the slag also melted and separated and rose to the top (because it was lighter), leaving molten iron containing a little carbon, but still with its original quantities of impurities, underneath. The carbon content could be adjusted to give the desired strength and hardness and the resulting steel could be poured off, free from slag, into moulds.

Crucible steel was expensive, partly because it was made from the already expensive wrought iron, and moreover, since no purification or control of impurities – other than slag – took place, its quality was variable. Even so, it was generally cheaper and better than most of the smith-made 'steel' used in swords and it was used extensively in small quantities for tools of high quality.

Crucible steel had the advantage that, instead of making a tool or a weapon dead hard outside and soft inside, any required strength and hardness could be achieved right through. For some purposes however the steel was still carburized or 'case-hardened' to give a more durable cutting edge, as indeed is still sometimes done. Nowadays nobody would try to make plain carbon steel in this way, except experimentally, but small batches of expensive alloy steels are usually made in crucibles.

Bessemer or Mushet steel

Until the middle of the nineteenth century crucible steel remained the only steel available and the metals of large scale construction were still cast iron and puddled wrought iron. The production of cheap steel in large quantities was originally due to Henry Bessemer (1813–98) and Robert Mushet (1811–91). Bessemer was a prolific inventor with a strongly developed business sense. After doing well with a number of inventions such as the manufacture of 'gold' paint and the consolidation of graphite for 'lead' pencils, Bessemer was drawn to steelmaking by the publicity given to the weaknesses of cast iron guns during the Crimean war.

After various experiments Bessemer had the revolutionary idea of blowing air through liquid pig-iron to remove the excess carbon and other impurities. He took out his master patents in 1855 but, in its original form, Bessemer steel was of poor quality because it contained excessive oxides and sulphur. However, in 1856 Mushet took out his own series of patents for a very similar process which differed from Bessemer's mainly in that the impurities which had not been burnt out by the air blast were controlled by the addition of something called 'spiegeleisen', a special cast iron from Germany containing manganese. It was the addition of manganese at the end of the blowing operation which ensured the success of the Bessemer process.

Bessemer steel is made in a contrivance called a Bessemer converter which consists of a pear-shaped crucible or container mounted on trunnions so that it can be tipped. It has no external means of heating. At the bottom of the converter are a series of holes or tuyeres through which air can be blown.

To operate the converter, the vessel is tilted so that its spout is under the outlet from a blast furnace and a charge of from five to thirty tons of molten pig-iron (that is iron at perhaps 1,200° C. containing about 4½ per cent of carbon and small amounts of silicon and manganese) is poured in. Since the converter is on its side the charge lies in the belly of the vessel and does not block the tuyere holes. Air is now blown through these holes and the converter is allowed to swing upright so that the air is forced to bubble through the molten iron. Under the conditions in the converter the air first oxidizes the manganese and silicon in the iron to form a slag which floats on top of the charge. The process is traditionally controlled by watching the colour and character of the flames produced in the mouth of the converter. At this stage they are short and reddish brown.

After a few minutes the manganese and silicon are fully oxidized and the air begins to remove the carbon; at this stage the flame changes to whitish yellow and becomes longer and more alarming. Finally, when all the carbon is eliminated, the flame drops and the blast is turned off. During the blowing period the burning of the carbon, manganese and silicon, which together amount to about six per cent of the charge, produces a great deal of heat; enough not merely to raise the temperature of the charge to keep pace with the increase of melting point due to loss of carbon, but so much that it would be overheated were not a little scrap steel added to cool it – otherwise the furnace refractories would be damaged.

By the end of the blow we have what is called 'blown metal' or approximately pure iron in the converter and it is usually desired to put back some carbon and manganese and also perhaps silicon. This is done by adding solid carbon and the special form of cast iron – spiegeleisen – which contains a high proportion of these elements. Manganese is wanted in steel both for its own sake as an alloying element and also because it controls the sulphur which is not removed from the iron during the Bessemer process.

Sulphur is a great nuisance in steelmaking because it does not oxidize to SO_2, as one might expect, but forms iron sulphide, FeS, and this has the peculiarity that it is soluble in molten iron but not in solidified iron. As a result, on cooling, the iron sulphide separates out at the crystal boundaries and weakens the steel (Chapter 4). The addition of manganese changes FeS to MnS, which is insoluble in liquid steel and so passes into the slag. Manganese also reduces the solubility of oxygen in steel which again is beneficial because oxide particles also tend to end up at the crystal boundaries.

Bessemer described his process to a meeting of steelmakers in 1856 in a paper called 'On the manufacture of malleable iron and steel without fuel'. Such was the enthusiasm of the audience and such was Bessemer's reputation that £27,000 was immediately subscribed in advance patent royalties, after which the steelmakers went home to build themselves Bessemer converters.

As it turned out none of them could make satisfactory steel, for the original Bessemer process was sensitive to the kind of pig-iron which is used and moreover it requires some skill. Not unnaturally, Bessemer became exceedingly unpopular. He then built a complete working model of his converter in his laboratory in St Pancras and used it to demonstrate steelmaking to his licensees. As these demonstrations did more to show up the incompetence of the licensees than to improve Bessemer's popularity, the agitation against Bessemer continued and no steel-maker would operate his process. In consequence, in 1859 Bessemer set up his own steelworks in Sheffield and made steel which sold extremely well, notably to the French and Prussian governments for the manufacture of guns.

The palpable success of Bessemer steel led steelmakers all over the world to seek licences and Bessemer received about one million pounds in patent royalties over and above the profits of his own steelworks. Those of Mushet's patents which were connected with blowing air through iron were probably anticipated by a few months by Bessemer but Mushet's patent on 'spiegeleisen' should have been valid and immensely valuable. Unfortunately Mushet forgot to pay the stamp duty when renewing the patent and thus extinguished his legal rights. In consequence Bessemer always refused to pay any royalties to Mushet and a long quarrel ensured. Towards the end of his life Bessemer paid Mushet an annuity of £300 a year.

It is difficult to quote representative figures to illustrate the economic effects of Bessemer steel. However, while in the 1850s steel fetched from £50 up to £100 per ton, by 1900 steel rails were being sold for less than £5 per ton. Although shipbuilders' tons are different from anybody else's it was possible to build steamships in the 1890s for around £10 per ton. (Nowadays, in spite of technical improvements, the price of steel is around £50 a ton and ships cost upwards of £200 a ton.)

Part of Sir Henry Bessemer's fortune went into building a cross-channel steamer, the *Bessemer*, in which the first-class saloon – which was large and luxurious – was hung, like a Bessemer converter, on trunnions and was intended to remain level at sea. Seasickness was further discouraged by blowing fresh air vigorously among the passengers by means of an ingenious arrangement of pipes in the floor.

In practice, the swinging cabin never worked properly and after a few voyages the ship was scrapped. The original saloon of the *Bessemer*, firmly fixed, is still in existence, as a conservatory, in a garden somewhere near Dover.

Nowadays, although straight Bessemer steels probably only account for a few per cent of the gross steel production, modernized and more sophisticated versions of the Bessemer system are making headway. In the Kaldo process, for instance, the converter is blown with oxygen instead of air and the extra heat which is produced is used to melt a flux for removing the sulphur directly and also to consume more scrap. What is gained in extra efficiency in such processes can however be lost by damage to furnace refractories and by the cost of the oxygen.

Open-hearth or Siemens-Martin steel

In a sense the Bessemer process was a cause of its own obsolescence because, as steel became cheaper and commoner, so did steel scrap and the availability of scrap began to exert an important influence on the economics of steelmaking. Nowadays about fifty per cent of the steel manufactured finds its way back to the steelworks as scrap The Bessemer process, however, in its traditional form is essentially a process for converting pig-iron from the blast furnace into steel and it only makes use of scrap in trivial quantities so as to get rid of a small excess of heat.

In the open-hearth process much of the charge consists of scrap steel which has the advantage, not only of being cheap in itself, but of already having most of the excess carbon and other impurities removed. Moreover it is rusty. As we have said, the Bessemer converter cannot use very much scrap because to melt it would need more heat than is produced by blowing the charge.

In 1856 the German-born Frederick Siemens (1826–1904) and his brother Charles William Siemens (1823–83), both of whom, like Bessemer, were prolific inventors with a strong business sense, developed the regenerative furnace. This is a furnace whose outlet and inlet are reversible and are both furnished with extensive labyrinths or honeycombs of firebrick through which the gases have to pass. Much of the heat in the outgoing gases is therefore stored in the outlet honeycomb. The furnace, however, is arranged so that the gas flow is repeatedly reversed and thus the incoming air is always being drawn over heated bricks and so picks up some of the heat which was in the exhaust gases. These furnaces are usually gas-fired and, as a consequence of their design, the temperature could be raised as high as the refractories would stand, in practice to something over 1,500° C. which was sufficient to melt pure iron.

In steelmaking the Siemens furnace was originally used simply as a convenient and economical way of melting crucible steel. Later, the Siemens brothers applied the regenerative principle to the traditional puddling furnace and made steel by melting pig-iron with iron ore. The addition of large quantities of scrap is due to a Frenchman, Pierre Martin, and dates from 1864.

In operation, the open-hearth furnace is charged with a flux – usually limestone – about equal quantities of scrap steel and pig-iron and some iron ore, perhaps Fe_2O_3. On heating, the whole charge is melted and the iron ore removes the carbon present in the pig-iron. The flux converts not only the non-ferrous oxides present in the iron ore to slag but also the sulphur present in the steel. For this reason it may not be necessary to add manganese. One of the advantages of the open hearth process as compared with the Bessemer converter is that much closer control is possible over the composition of the steel and until recently about eighty-five per cent of plain carbon steels were made in this way. However the open hearth furnace is slow and rather expensive to operate and it has therefore been losing ground to the oxygen blown converter processes such as the Kaldo which are quicker, rather cheaper and can take liquid iron directly from the blast furnaces. In consequence only about half the steel in the country is now made by the open-hearth process which seems to be declining rather rapidly.

For still closer control over cleanliness and composition it is usual to employ electric furnaces and a comparatively small, but significant, tonnage of high grade steel is made in this way.

Chapter 11   The materials of the future

or how to have second thoughts

> *'In fact we have to give up taking things for granted, even the apparently simple things. We have to learn to understand nature and not merely to observe it and endure what it imposes on us. Stupidity, from being an amiable individual defect, has become a social vice.'*

J. D. Bernal, *New Scientist,* 5 January 1967.

Before the first edition of this book was published in 1968 I was persuaded, rather against my judgement, to add a final chapter called 'The materials of the future' with the defensive sub-title 'or how to guess wrongly'. Much to my relief there do not seem to be any real technical clangers in the original chapter and, in the present edition, it has been possible to transfer most of these prophecies, as accomplished factual information, to their appropriate places within the body of the book.

The weakness lay, I think, not in the facts but in the emphasis. Much of the discussion was about what it is now fashionable to call 'High Technology' and we are less inclined to be impressed by this kind of thing than we were a few years ago. Getting to the Moon has turned out to be, perhaps, just a little boring and that enterprise was, in any case, a very expensive way of developing non-stick frying pans. Students tell me that space-fiction keeps its popularity but surely its appeal is no longer chiefly technical, it is simply the fairy-story or the fable in modern dress.

More and more one comes to see that it is the everyday things which are interesting, important and intellectually difficult. Furthermore the materials which we use for everyday purposes influence our whole culture, economy and politics far more deeply than we are inclined to admit; this is, indeed, recognized by the archaeologists when they talk about the 'stone age', the 'bronze age' and the 'iron age'. It is significant that, from very soon after the introduction of bronze for tools and weapons, there was a sort of polarization between the metallic and the non-metallic technologies.

Although the ductile metals have their limitations and they are by no means infallibly tough or even particularly strong yet, for a number of purposes in peace and war, they were so much superior to the other solids which were available that a whole range of popular beliefs and superstitions and emotions have become associated with them and especially with iron and steel. In one shape or another these emotional situations have continued down to the present day and, because they still influence many people's thinking, it seems to me that it may be worth spending a little time on the historical and psychological background of the subject.

Because of their interest in tools and weapons the early civilizations were inclined to have smith-gods. The Norsemen had Thor and the Greeks had Hephaistos, who is also the Roman Vulcan after whom volcanoes are called. The stories about Hephaistos are many and not always edifying. Hephaistos, be it noted, was an Olympian god and the Olympians were not indigenous to Greece but came, as it is sometimes said, with the Dorian invaders from the North, who may have conquered the Greek world with iron weapons – possibly at the fall of the Minoan kingdoms around 1200 B.C. At any rate Hephaistos looked after the supply of iron and of weapons and was regarded by gods and men with rather mixed feelings. Hephaistos was rather a big-wig as gods went and he has a magnificent temple in Athens which stands to this day. It is interesting that there was no god of any consequence specifically connected with any of the non-metallic materials; one or two scruffy little low-class demons were associated with pottery works – where they sometimes caused the pots to crack in the kilns – but these hobgoblins were of no real importance in the world.

In the middle ages Wayland Smith, who is Wieland, who is Voland and who lived in Heligoland, or possibly near the White Horse in Berkshire, was continually turning up in improbable circumstances. There were magic swords almost innumerable, Miming, Flamberge, Excaliber and so on; some of these were quenched in Dragon's Blood. If he was not too busy with a magic sword Wayland Smith would shoe your horse for you for a very modest sum. Again, I do not think that there was any real non-metallic equivalent to Wayland.

Although this sort of thing died out – or was suppressed by the Church – the strength of the feeling about the special nature of iron and steel continued and was reinforced rather than diminished by the Industrial Revolution. The Victorians could not really be said to have had smith-gods but they did everything they could with iron – short of affording it actual formal worship – and iron and steel became synonomous with industrialism and technical progress. Everything done with metal was important and to be taken seriously; it followed that anything done with non-metals was old-fashioned and quaint and unimportant.

About 1912 an ex-theological student called Joseph Djugashvili (1878–1953) changed his name to Stalin* since 'Iron out of Calvary is master of men all'. He proceeded to live up to his name and, incidentally, spent much time later on in building up the steel industry throughout the Soviet Empire. Actually it looks as if he may have rather overdone it since most Communist countries now seem to be suffering from a surplus of steel.

Of course, from Boulton and Watt down to Henry Ford, iron and steel have been, above all other materials, the agents of development, industrialization, enrichment and 'progress' and, historically, we could not have done without them; the question is whether we ought now to be growing out of that kind of thing. Everything has its bad side and the bad side of metal technology did not come to an end with the Factory Acts. What really made the industries which were based on iron and steel so successful in modern times was that they lent themselves best to disciplined mass-production and to the dilution or elimination of skill and individual judgement at the bench level. All the worthwhile decisions were taken at the top and imposed from the top. Such manufacturing systems placed enormous power in the hands of managers and accountants and, eventually, in the hands of governments and trades unions. Governments and trades unions are no more enthusiastic about giving up this power than the managers and accountants and they are even more apt to talk about the economies of scale and the benefits of 'rationalization' and centralized control.

* Both the English and the Russian words are supposed to derive from the Old Teutonic 'stah' or 'stag', to be firm or rigid.

What is left out of account is that most ordinary people don't like this kind of thing and nowadays they find ways of not standing for it and of being bloody-minded which more than nullify the economies of scale. A factory may be an accountant's paradise but if it is on strike it is not producing anything at all and surely it would be better to have a less 'efficient' system which did produce something?

These large and highly interdependent organizations presuppose a willing and disciplined labour force of a kind which is most unlikely to exist for long under any political system. In my opinion nationalization and high wages or even 'workers' control' do not really improve the situation because they do not deal with the roots of the trouble which are to a large extent inherent in the size and in the inhuman nature of so many manufacturing processes.

The difficulty has been that, until recently, no economically practicable alternative to bigness has existed and, on the whole, the economists from Marx to Keynes seem to have taken the economies of scale for granted; indeed most of them have rather welcomed bigness, partly because it appeals to tidy minds, but also because bigness puts more power into the hands of the Children of Light. But the Children of Light are few in this world and people furiously disagree as to who they are; also all power corrupts and ought to be diminished.

But then what shall we do to be saved? There is nothing like Leather, says the cobbler; nothing like Sound Management, says the business man; nothing like Taxation, says the politician; nothing like Equality, says the trades unionist; but I think that the problem might very well be a technical one. Although about twice as much timber is used as steel, it is interesting that the woodworking industries in the Western world are seldom in the news, whether for good or for bad and they do not seem to be especially plagued with troubles or dissention. Somehow steel and its dependent industries is never out of the headlines and nobody associated with it ever seems to be really happy. Whether we are concerned with making swords or making motor-cars steel somehow remains the material of power and politics. It has been the fashion, especially on the Left, to regard steel as being in some way a 'People's material', a key to all sorts of enlightenment and progress. Of course steel is extraordinarily useful stuff, we cannot get on without it and it is wrong to take too simple a view, yet there is a strong case that steel is the agent of a sort of faceless industrial oppression, the life-blood of the Dark Satanic Mills. Indeed steelworks are gloomy places.

The differences between the social records of the industries based on wood and those based on metals is unlikely to be due to pure chance and can probably be traced, however indirectly, to the physical characteristics and properties of the materials themselves. For one thing, speaking comparatively and in a very general kind of way, people seem to *like* working with timber; for another, the very manufacturing limitations of wood impose constraints upon the sociology of the industry which may well be salutary. If this is true, then it might be a good idea to study the whole question of how materials affect the happiness of the people who work with them in much more depth. Now that we can design new industrial materials to suit ourselves we ought surely to bear this psychological element in mind much more than we do. At present nobody seems to care.

Designing materials to suit ourselves

All this brings us to the question of designing new materials and we might begin by harking back just a little. Materials science – the study of materials as a whole rather than in their special chemical, physical and engineering aspects – is a fairly recent development. Indeed it has only lately become respectable. However, in spite of its youth, the new discipline has been rather successful and I think it is fair to say that we now understand a great deal more about the reasons for the mechanical behaviour of solids than we did only a very few years ago. This may be because many of the raw materials of understanding were lying to hand already. There was a great body of orthodox physical and chemical knowledge and there was also, although in different hands, an accumulated mass of engineering experience and tradition. To fit them together and to make one explain and confirm the other required only a moderate amount of original experiment and fresh thinking – once a sufficient number of workers had considered it worthy of their serious attention. As so often, the real difficulty about the solution of a problem is to recognize that the problem exists.

Naturally enough, the first task was one of understanding the observed phenomena – why solids in general, and especially the familiar materials, behave in the way they do. Although there are still a good many loose ends, this stage can broadly be said to be accomplished. The problem now facing materials scientists is what use to make of their knowledge. The possibilities are not unlimited – one of the things which has been learnt is that there are a considerable number of things which cannot be done even if we wanted to do them. A cognate aspect of our knowledge is that many lines of improvement had already been exploited, almost to the full, before the scientists got there. Some of our knowledge is useful simply in telling the engineer what to avoid – what sort of stress-concentrations are dangerous, for instance.

However, when all due reservations are made, the ambitious will want to apply materials science in radical ways, either by making substantial changes in the older materials or else by inventing new and perhaps better ones. However, the more intelligently we examine traditional materials like wood and steel the more we ought to be impressed by how cunningly they are made. What is wanted is not one property in isolation but rather a balanced combination of properties and this is provided in such materials by means of exceedingly subtle and complex mechanisms such as dislocations and hollow helical tubes. Modifications to the older materials of construction are more likely to spoil them than to improve them.

If we are going to set out to invent entirely new materials then we had better watch our step because the requirements for any really successful material are likely to be very complex indeed. Nevertheless, it is probably worth trying and the eventual rewards, both economically and socially, may be very great. As Academician Rabotnov remarked in his introduction to the Russian translation of this book 'this is a noble task to which young people might well devote their lives'. Although I might not have expressed the sentiment in quite those words, I cannot but agree with Professor Rabotnov.

Before we can discuss the subject of new materials intelligently we have to begin by asking ourselves the question – 'What do we really mean by a *better* material – better for what?' The answer to this is anything but obvious and the question is really the central one in materials science at the moment. As we have already said, the whole problem is made more difficult by the extraordinarily complicated way in which the technical, social and economic aspects are interwoven.

Many business men however seem to have no serious doubts about what they want – theirs is a simple Faith and they want the stuff to be cheaper. For this reason a high proportion of materials research is directed towards getting the cost of production down. There are, of course, some materials which are unduly expensive and which could and ought to be made more cheaply. However, I do not think that this is true to any important extent of the engineering materials of construction in general, such as steel. The cost of steel was reduced very greatly during the nineteenth century and it is now fairly cheap; on technical grounds it seems unlikely that such marginal economies as are likely to be made in future can really justify the investment of any large scientific effort.

Even if it were practicable to achieve large reductions in the cost of constructional materials it is worth asking who would benefit. In many finished products the cost of the materials, as such, is only a few per cent of the cost of the finished article and so, even if the material were to be supplied free, the benefit to the consumer would be small compared with other changes which might be introduced.

Secondly, the very cheapness of materials may actually have a bad effect upon design and performance. When materials are unduly cheap the designer has too little incentive to economize in using them and the result may not only be heavy, clumsy design (involving waste of fuel and damage to roads) but it may also, in the long run, lower the professional standards of engineers. The reason for much bad, heavy, ugly design is partly to save fabricating costs but as often, I am afraid to save the designer from the trouble of thinking. Economists tell us that, in a given context, there is generally an optimum price for both land and labour which results in efficient social development and I fancy that the same reasoning might apply to the price of materials.*

There is another aspect of the price of materials which is also important. The cost of the various fabricating processes is frequently very many times higher than the cost of the bare material. For instance it is cheaper to buy a plastic at 20p a pound which can be moulded into its final shape for a further 5p than to buy steel at 5p per pound which needs to have 50–100p spent on it in pressing, machining and finishing operations.

Cheap fabrication is, at root, the reason for the commercial success of plastics which are always expensive raw materials compared to most metals. It is not only that, from the economic point of view, such materials can offer large savings in processing, manufacturing and finishing but also many of these processes can be carried on competitively upon a very small scale. I have seen the extrusion of plastic tubing carried on as a village industry in Hungary and indeed, all that is needed is two or three extrusion machines and a shed to house them. It is difficult to see how the process could benefit, economically or otherwise from being done in a big factory. When I was on the Board of a large plastics company I was able to watch our extrusion business being run off the market by little back-street firms whose overheads were lower and whose thought-processes were quicker than ours.

Very much the same sort of thing applies to the injection-moulding of thermoplastics; an industry which makes millions upon millions of toys and kitchen gadgets and useful what-nots at an incredibly low price. Given a suitable (and rather expensive) mould, injection moulding can be carried out in almost any barn

---

* During the nineteenth century Edward Gibbon Wakefield (1796–1862) pointed out that an important obstacle to the colonization of Australia and New Zealand was the very low price of land. This, he showed, led to so wasteful a use of land that no colony could flourish. When, at his instigation, the price of unoccupied land was controlled and raised colonization was successful. There are many books on Wakefield.

or garage. It may suit the convenience or the vanity of administrators or business men to group these activities in large factories but there does not seem to be any compelling reason for doing so. Though I find it difficult to go all the way, politically, with either William Morris or Dr Schumacher, there can surely be no doubt at all that Small is Beautiful?

Extrusion and injection-moulding are very well as far as they go and they enable jobs which are inherently dull and repetitive to be carried on by smaller groups of people in pleasanter surroundings – which is all to the good and, after all, some people like repetitive jobs – but, of course, it does not give much scope for initiative and variety in the design of the product.

This is achieved, to a considerable extent, in the wood-working industries, making products like furniture in comparatively small numbers and to comparatively individual designs. By using modern machinery and adhesives and finishes the cost can be kept competitive with the big factories. The same thing applies to a still greater extent in the fibre-glass industry where all sorts of shell-like products, from swimming-pools to boats to car bodies to flower-pots, can be produced in little factories with very low capital costs.

Naturally, there are plenty of small industries which are based on metals but these people generally find it more difficult to compete with the big boys and it is the non-metallic small industries which are going ahead much the faster. This is probably an area where a moderate amount of materials research could do a great deal of good.

Energy conservation

Latterly it has become very fashionable to talk about energy conservation and, quite apart from political questions about oil-wells, it is probably high time that we gave this aspect of technology more consideration and respect. When power had to be provided, in very small quantities, by men or horses, vehicles were generally made from wood and other light materials. It is most instructive to look at sedan chairs and horse carriages in museums and to see how light and robust some of these things were.

Traditionally, railway trains were drawn along smooth and level rails by engines burning lots of cheap coal and a railway carriage is an interesting example of just how heavy and clumsy one can make a moving structure. To some extent the prodigal use of material in railway rolling-stock arises from the nature of trains themselves; that is to say trucks and carriages have to be constructed to resist what are called 'buffing loads' which occur because of the mutual impact of strings of vehicles during shunting and manoeuvring. Light, modern rolling-stock has usually to be worked in conjunction with older, heavy vehicles which are liable to smash anything near them and, in my experience, it is difficult to save very much weight or fuel in trains by advanced structural design. In any case, I doubt if the real problems of the railways are the technical ones.

Nevertheless, the whole tone of traditional engineering was set for many years by railway and by marine engineering and by bridges and stationary engines and all these things were generally constructed as if weight were no object. 'The English think that weight means strength.' The requirement for weight saving arose with the invention of the motor-car and the aeroplane but even here engineers were generally more interested in increasing the power of the engine than in saving weight in the structure. It used to be said that 'a tea-tray will fly if only you put enough power into it'.

By running internal combustion engines at very high speeds on high-octane fuel it was possible to increase their power output very dramatically indeed. Much of this power output has been needed to cope with the large increase which has taken place in the tare weight of cars during the last fifty years, mainly due to the introduction of the pressed steel body. As long as oil was comparatively cheap and plentiful I suppose that this did not matter very much. In fact the fuel consumption of a car is roughly proportional to its weight and so nearly half the petrol which is imported into this country is used or wasted in propelling heavy rusty steel shells along the roads.

The only real argument in favour of pressed steel bodies for cars is the safety one. These steel shells are in fact very good at absorbing impact energy during an accident – due to the high work of fracture of mild steel – and so each of us drives about inside his own private tank. However, according to my sum, the extra fuel needed to move all this armour around costs about £1,000, at present prices, taken over the life of each average car. One would think that there would be cheaper forms of protection than that. The total cost to the country in imported oil is enormous and, of course, the extra weight does no good to the tyres or the roads.

Fibre-glass car bodies save a great deal of weight and, in fact, they are already quite widely used. The trouble is that such bodies have barely enough energy absorption and they are probably distinctly less safe than steel bodies. What is needed is a large increase in the work of fracture of the material and also, perhaps, some increase in stiffness. As we saw in Chapter 8, tougher, stiffer composites seem to be only just around the corner and, if so, we may see a large swing to plastic car bodies together with a saving in fuel which might well be as high as 30 per cent. Since such bodies do not rust cars would probably last longer.

Light-weight car bodies are one good way of saving energy; improved building methods and house insulation are probably another. Unfortunately development in building materials is a good deal handicapped in this country by conservative building regulations and by the Building Societies. It does, however, very much look as if a new approach to the whole question of building materials and insulation is overdue.

The new materials might perhaps be sophisticated ones based on polymers or other 'advanced' substances but they might equally, and perhaps preferably, be based on traditional local materials, modified where necessary in the light of modern knowledge. Professor Biggs tells me that if one looks at housing from the total energy point of view – that is to say, if one considers the energy needed to make and to transport the materials, as well as the energy consumed during the life of the building in heating it – then the traditional English thatched and half-timbered cottage is probably better than anything else. Rather predictably, this is just what one is not allowed to build under the present regulations.

A variant of the same problem exists in connection with overseas developments. In many tropical regions, such as parts of Africa, it is difficult to improve upon the traditional construction of mud and reeds which can be put up cheaply and quickly and which is cool to live in. Most of the drawbacks of these buildings can be got over by making use of various forms of modern technology. I am told however that the inhabitants reject this construction with scorn and insist on building with concrete and corrugated iron which are both hot and expensive. As long as such attitudes prevail it is difficult to see what science can be expected to do.

So far we have talked about saving energy; however, if we are going to go out and actually collect energy from primary sources, such as the sun and the wind, then we shall probably need to invent and to make use of a whole new range of materials. For something like 200 years engineering 'progress' has consisted of making machinery and other engineering devices more and more compact, of working at higher and higher stresses and of 'processing' more energy in less space. A modern ten horsepower engine is at least a hundredfold smaller and lighter than it would have been in the eighteenth century. This has been possible because of the concentrated nature of fossil fuels. Primary energy sources such as the sun and the wind are not concentrated and have to be collected in driblets, as it were, from over a wide area. Plants are superbly good at collecting the energy of sunshine but the structures and materials which they use are different from those which are popular with the modern engineer.

If we want to revive the sailing ship and the windmill, under modern conditions, as I am fairly sure that we should, then we shall have to devise ways of reaching out and collecting small amounts of kinetic energy from quite long distances very cheaply and simply. One has only to sit down and study the problem from the economic and structural point of view to see that this probably cannot be done with existing materials and we shall probably have to go about the job in some more intelligent way and, again, this will undoubtedly call for new kinds of materials and structures.

The disposable technologies

There is no doubt at all that the introduction of 'disposable' materials, such as cheap paper and plastics, has been, on the whole, decidedly beneficial. It has reduced the incidence of infectious disease and dramatically diminished the less attractive kinds of domestic chores. The question is, how far ought we to go in this direction? What do we really mean by 'disposable' anyway?

It is all very well to throw away things like towelling and packaging and it is probably an excellent thing to throw away your morning paper, but what about refrigerators and motor-cars? It is perhaps possible – just possible – to exaggerate the cynicism of manufacturers and shopkeepers about 'planned obsolescence'. Of course, Nature arranges for all her creatures to die and there is no good reason why the works of man should last for ever. A paper towel is very cheap – much cheaper than one made from cotton or linen – and furthermore the cost of washing a durable towel is probably greater than that of throwing away many paper ones. But all this is not true of motor-cars or refrigerators. In such things a small addition to the first cost would result in a much longer life and in reduced maintenance costs.

One of the biggest contributions to the total economy would be to make 'consumer durables' durable. After all, if such things last twice as long, that is roughly equivalent to doubling the production of the article; in other words one need only make half as many. This does not necessarily cause unemployment because one can use the money which is saved to buy something else – or else perhaps just save it.

Many products wear out – or rather go out of service – for silly, shoddy reasons (Oh no, sir, we don't stock spares for out-of-date models). However, the biggest single reason for throwing things away is probably rust. As we have seen the introduction of plastics in place of pressed steel shells is one way of getting over this and very often these plastic shells can be manufactured on a 'Small is Beautiful' basis.

I suppose that industries which deliberately live by 'planned obsolescence' ought not to be surprised if they become obsolete themselves; perhaps we need some form of euthanasia for declining industries. At present the policy is to pension them off so that they can continue to exist comfortably at the taxpayer's expense. One does see that there are political and social problems about lame ducks and disposable industries but it is seldom pointed out that it is precisely those industries who, in their hey-day, were most ruthless in superseding the stage-coachmen, the canal boatmen, the coastal seamen, the sailmakers, the millwrights and such other excellent people who now call most loudly for subsidies on social grounds. Times have changed since the nineteenth century but are they changing fast enough today?

If 'consumer durables' wear out too quickly, do houses and other buildings last too long? Very possibly more houses have been built during the last thirty years than were put up in the whole span of previous history; most of them are ugly, or at any rate commonplace, and our grandchildren may well want to get rid of them. Also it would perhaps be better if our houses could be altered quickly and cheaply to suit our changing requirements and better still if we could take them with us when we wanted to move to another part of the country.

Unfortunately the local authorities have joined with the Building Societies to ensure that houses shall be as expensive and as immoveable as it is possible to make them and also very difficult to modify. It is quite true that both caravans and also most temporary buildings, in their present forms, are nearly always ugly and shoddy and to be discouraged. But is this necessarily so? We have got it firmly into our heads that anything temporary or disposable has to be ugly but, after all, one could hardly have anything more temporary or more disposable than a crocus or a daffodil and it might be worth giving some thought to the way in which Nature does these things.

We have, I suppose, the biggest aesthetic opportunity in the whole of history and what are we doing with it? Nothing, or rather we are producing miles and miles and miles of the dull and the commonplace. Does everything really have to be Fabian grey? By making use of modern ideas and modern materials I suppose that we could all of us be pavilioned in splendour – quite comfortably and cheaply – if we really gave our minds to it.

Because the influence of materials and structures upon almost every aspect of our lives is so great, the fact that we now at last understand, pretty well, how materials work and have some idea of how to invent new ones, is very important indeed. So far the implications of this new knowledge have not been sufficiently widely appreciated. Our modern understanding of materials has been brought about by getting engineers and physicists and chemists to talk to each other, which they were rather reluctant to do. What is wanted now, I think, is to bring materials science together with economics and aesthetics. If this can be done in a really imaginative way then the opportunities are enormous and surely the challenge is great enough to satisfy the most able and the most ambitious?

Appendix 1 On the various kinds of solids

and what about treacle?

Atoms and molecules

There are roughly a hundred different kinds of atoms, not counting isotopes (that is atoms with the same chemical properties but slightly different nuclei.) About ten of these are ephemeral things which have been made by atomic scientists by artificially transmuting other atoms. The rest have existed naturally, almost since the beginning of time, and usually in a more or less immutable state. Of these only about twenty or thirty are sufficiently common to concern us in this book. Each kind of atom is called an element. All the atoms of any one element are virtually identical but the atoms of the different common elements differ greatly in their properties. This is basically why matter displays the enormous variety which it does.

The atoms of an element can exist in a pure, homogeneous state as elements, e.g. iron (Fe) or carbon (C); they may exist as any kind of mixture or solution (whose properties are as a rule simply a mixture of those of the constituents) or as definite chemical compounds, e.g. iron carbide, Fe_3C. In this last case the atoms usually combine in quite definite fixed proportions to yield a chemical compound which is an entirely new substance with its own characteristic properties. For instance chlorine (Cl) is normally a green poisonous gas; sodium (Na) is a soft shiny metal. Equal numbers of atoms of each combine to form sodium chloride, NaCl, a white harmless powder, ordinary table salt.

The twenty or thirty common elements can and do combine to form a huge variety of substances, solid, liquid and gaseous and this, of course, is what chemistry is about. The four common elements, carbon, hydrogen, oxygen and nitrogen can combine in virtually infinite variety to form nearly all the substances which are found in living matter besides a great number of non-living and artificial compounds such as plastics, petrol and oils, drugs, paints and so on. Compounds of this type are called 'organic'. Other compounds are lumped together under the label 'inorganic'.

This includes most compounds of metals such as minerals (other than oil), ceramics and so forth.

In the game of atomic Meccano which is called chemistry the basic unit is the molecule, that is the smallest particle of a compound possessing the properties of that compound. For instance a molecule of sodium chloride is NaCl, that is, one atom each of sodium and chlorine. A molecule of benzene is C_6H_6, that is a construction containing six atoms of carbon and six of hydrogen. Again, a good many elements exist in the form of molecules, combined so to speak with themselves. Iodine, for example, exists as I_2. It will be realized that the word 'molecule' is used rather loosely to describe a small grouping of atoms according to chemical principles. This does not mean however that the chemical bonds which hold some molecules together are not sometimes very strong indeed. Molecules vary in size from little things made from a couple of atoms up to elaborate structures made by joining together hundreds and sometimes thousands of atoms.

The biggest organic molecules can be relatively quite large, perhaps several hundred Ångströms long. Inorganic molecules are generally, though not always, smaller, typically perhaps about ten Ångströms. However there do exist long inorganic chain molecules, such as asbestos, which are quite as long as anything organic.

Chemical bonds

Matter, that is atoms and molecules, is held together by chemical bonds. These are of several types which vary a good deal in their properties.

(a) BONDS WHICH USUALLY HOLD ATOMS TOGETHER WITHIN THE MOLECULE

Covalent bonds These occur when two atoms share a pair of electrons. It is usually the hardest bond to make but, when made, it is often very strong and rigid. This is the type of bond that occurs within organic molecules and sometimes in ceramics.

Ionic bonds The individual atoms of elements are, as a whole, electrically neutral since the charges of their constituent particles balance out. When sodium reacts with chlorine the metal gives one outer electron to the gas so that the sodium is now positively charged and the chlorine negatively charged. As a result the two atoms now attract each other. In solids, ionic bonding is particularly common with metallic compounds. It is quite common for the bonds in some compounds to have a composite ionic and covalent character. Whereas covalent bonds are strongly directional, ionic bonds operate more uniformly in space around the charged atoms.

Metallic bonds These are the bonds which, in general, hold together metals and their alloys when these elements are not in the form of a definite chemical compound. In this case, some of the outer electrons are not held permanently in orbits related to particular atoms but can rove freely through the material being, as it were, the communal property of all the atoms in that piece of metal. They form what is sometimes known as an 'electron sea'. Metallic bonds are comparatively easy to form, to break and to reform. Because of the free movement of electrons within the material, metals are good conductors of electricity.

(b) BONDS WHICH USUALLY ATTACH MOLECULES TO EACH OTHER

Hydrogen bonds Although the water molecule, H_2O, for instance, is, as a whole, electrically neutral, the distribution of the charges within it is such that there is a strong local charge imbalance which can be highly attractive. Many organic compounds, in particular, have along their molecules numerous $-OH$ groups each of which is capable of attracting other $-OH$ groups, water and so on. It is these forces which, in general, keep the molecules of plants and animals attached to each other. These $-OH$ groups are often called 'Hydroxyls'.

van der Waal forces These are much weaker forces which arise from small local variations of charge which tend to occur all over the surfaces of any molecule and are not generally associated with any particular chemical groups. Cohesion depending upon van der Waal forces is not particularly common in nature but is quite frequent in plastics and between other artificial organic molecules. Because of van der Waal forces, almost anything will stick to almost anything else in rather a feeble sort of way provided that the surfaces are sufficiently clean and in good contact.

Heat and melting

Although temperature can rise as high as you like, it cannot fall below absolute zero which, as Kelvin calculated, occurs at $-273°$ Centigrade (which is not all that cold, considering that one can get temperatures of $+4,000°$ C. or more without much difficulty). At absolute zero all atoms and molecules of whatever substance are at rest and everything is a hard solid.

As soon as the temperature starts to rise, all the atoms and molecules begin to vibrate. The higher the temperature the more vigorous the movement, heat being, of course, simply the random vibrations of the atoms and molecules in a substance. Up to $0°$ C. water, for instance, remains in the form of a hard solid, ice, although its internal cohesion is subject to more and more battering from the thermal agitations of the molecules as the temperature rises. Finally, at $0°$ C., the internal cohesion of solid ice gives up the struggle and the molecules acquire the sort of shuffling freedom which characterizes a liquid. The ice has melted to form water. If the temperature rises further, the water molecules shuffle more and more energetically until at $100°$ C. the water boils to form a gas or vapour, steam, in which the molecules, released from any kind of mutual cohesion, fly about quite freely, like a swarm of bees. If the temperature is then lowered, the same sequence occurs in reverse so that the steam will condense and eventually freeze. These changes of state are usually considered as physical ones since the only bonds which have been broken and reformed are the hydrogen bonds between the water molecules which are, within themselves, unaffected.

Although the melting and boiling temperatures of various substances vary enormously, the general principles are very much the same for everything. A few substances (such as Iodine and smelling salts, ammonium chloride NH_4Cl) pass straight from solid to gas without the intervening state of liquidity. Quite a number of complicated solid compounds decompose chemically when heated before they get a chance actually to melt or vaporize.

Crystallization

As we have said, all molecules attract each other to a greater or less extent and, if they are not torn apart by thermal agitations, they will cohere into some kind of clotted solid. Since the molecules attract each other, they naturally try to get as close together as possible and this is to be achieved, not by huddling together in some hugger-mugger way, but by systematic packing. A loose heap of bricks occupies more volume than bricks properly stacked. In so far as the molecules have a more or less free choice of sites, therefore, they tend to take up regular positions, like bricks in a wall, when they solidify. This can occur when a liquid remains quite fluid up to the moment of freezing, as happens with water and metals for instance. In these conditions a molecule, looking for a desirable residence, will shuffle around for a long time till it finds a 'closely packed' position to sit down in.

Mechanisms of this kind lead to the formation of crystals which are essentially highly regular arrangements of molecules in a solid. In a good crystal the number of molecules and the regularity of their arrangement is breathtaking.

All solids have a tendency to be crystalline but not all achieve it. However, all metals and many simple inorganic substances are invariably crystalline. Other substances may be so to a greater or less extent. Good crystals are often quite large, up to several inches across, and may show characteristic simple geometrical shapes.

Glasses and ceramics

If a liquid is very viscous before it freezes, or if it is cooled very quickly, most of the molecules will not be able to select the most closely packed sites and the material will solidify in the form of a hard irregular mass of molecules. The most usual cause of high viscosity, when a liquid approaches its freezing point, is that the molecules, instead of remaining independent until they crystallize, associate while still in the liquid, to form sluggishly mobile chains or networks. This is particularly the case with some metallic oxides such as silica, SiO_2, that is to say sand, which melts to a viscous liquid and when subsequently cooled very readily forms a glass. The melting point of pure silica is inconveniently high (1,600° C.) and so most practical glasses consist of mixtures of sand, lime and soda which can be melted in ordinary furnaces.

Sugar, as bought at the grocer, is normally crystalline (and there is, of course, a special trade in large, well shaped crystals for coffee). However, if sugar is melted and cooled fairly quickly, it will form a glass, toffee. Treacle, of course, is a viscous liquid on the way to becoming a glass.

'Fudge' is devitrified toffee. That is toffee which has wholly or partly abandoned the glassy state and has crystallized. Because of the consequent shrinkages fudge is full of cracks and thus weaker than toffee (Chapter 4).

Metals and substances like sugar and salt are crystalline and so are the majority of rocks, which had plenty of time to cool. Glass, obsidian and toffee are glassy. Some materials are a mixture of the two. Most common ceramics, such as domestic pottery, consist of small crystals of metallic oxides and silicates which, when they are 'fired', become stuck together with a thin layer of glassy material. If the pottery is glazed this is done by melting on to the surface a layer of a low melting point glass.

Polymers

Most living matter, both animal and vegetable, and all plastics are in the form of polymers. Polymers are in many ways very convenient solids to make, both for Nature and for the plastics technologist. The basic idea is that convenient small unit molecules are built up, on the site as it were, into large chains or networks.

Elastomers

This includes things like natural and artificial rubbers. These are rather like ordinary polymers, in that their molecules consist of long chains, but these chains are not closely tied together laterally, so that the chains can fold and coil up. The enormous mechanical extension of rubber is due to the unfolding and uncoiling of these chains.

NOTE ON CONVERSION OF UNITS

Length 1 metre $= 3 \cdot 28$ feet
        1 foot $= 30 \cdot 48$ centimetres (exactly)
        1 micron (μm) $= 10^{-4}$ centimetre $= 3 \cdot 937 \times 10^{-5}$ in.
        1 nanometre $= 10$ Ångströms $= 10^{-9}$ metre

Area   1 square inch $= 6 \cdot 4516$ square centimetres exactly

Force 1 kg force $= 2 \cdot 205$ lb. force
                  $= 9 \cdot 81$ Newtons
                  $= 0 \cdot 98 \times 10^6$ dynes

Stress or pressure
             1 p.s.i. $= 6895$ Newtons per square metre
                    $= 0 \cdot 0705$ kg per square centimetre
                    $= 6 \cdot 9 \times 10^4$ dynes per square centimetre
        1 MN/m^2 $= 10^6$N/m^2
                    $= 146$ p.s.i.
    1 atmosphere $= 14 \cdot 696$ p.s.i.
                    $= 1 \cdot 033$ kg per square centimetre

Energy   1 Joule $= 10^7$ ergs
                  $= 0 \cdot 239$ calories
                  $= 0 \cdot 734$ ft. lb.

Appendix 2 Simple beam formulae

or do your own stressing

The basic formula for the state of affairs at any given point along a beam is:

$$\frac{M}{I} = \frac{s}{y} = \frac{E}{R}$$

Where $M =$ the *bending moment* to which the beam is subjected at this point. Bending moment is load multiplied by distance, e.g. in inch pounds. If M is not, for a given case, obvious from first principles it may be obtained from the data given below.

$I =$ the *moment of inertia* of the section of the beam at this point, e.g. in inches^4. Moments of inertia are treated in text-books of elementary mechanics. The moment of inertia in question here is that about the neutral axis which, in most beam problems, will be a line through the centre of gravity of the section. If the I of the section is not already known it is usually easily calculated by applying one of the following formulae.

For a rectangle:

$$I = \frac{bd^3}{12}$$

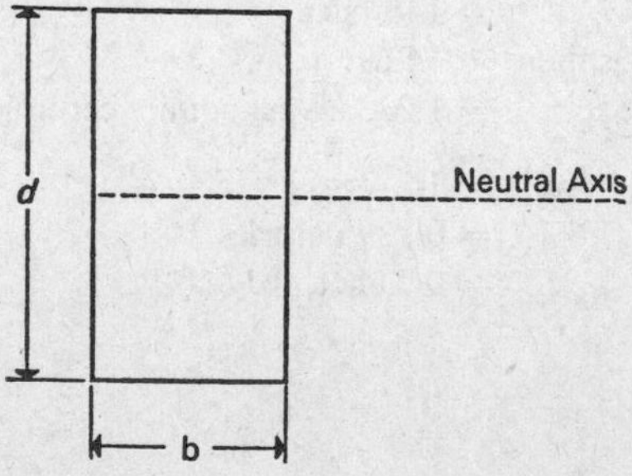

Neutral Axis

Figure 1.

Hence sections made up of rectangles, such as rectangular tubes, H beams and channels can often be calculated by subtracting the I of the empty areas.

For a circle:

$$I = \frac{\pi r^4}{4}$$

Figure 2.

Again hollow tubes can be calculated by subtraction.

$s =$ stress in material.

$y =$ distance from the neutral axis. In a symmetrical section this is the distance from the mid-point or centre of gravity of the section.

$E =$ Young's modulus.

$R =$ radius of curvature of beam when it bends under load.

Of these various properties we most often want to calculate the stress in the beam and so:

$$s = \frac{My}{I}$$

For a simple rectangular beam:

$$I = \frac{bd^3}{12} \quad \text{and} \quad y = \frac{d}{2}$$

at the surface where the stress is greatest. So:

$$s_{max} = 6\frac{M}{bd^2}$$

Which is why a beam twice as thick is four times as strong and so on.

For specific beams the following information is useful. Information covering virtually every imaginable form of beam is given in *Formulas for Stress and Strain* by R. J. Roark (McGraw-Hill, 1954) which should be referred to for more complicated cases.

Point loads

(a) Simple cantilever length l with point load W at end.

FOR BENDING MOMENT

at any point distant x from the end:

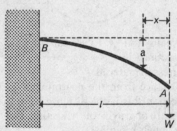

Figure 3.

$$M = Wx$$
$$\text{Max } M = Wl \text{ at } B$$

FOR DEFLECTION

at any point:

$$a = \frac{1}{6}\frac{W}{EI}(x^3 - 3l^2x + 2l^3)$$

$$\text{Max } a = \frac{1}{3}\frac{Wl^3}{EI} \text{ at } A$$

$$\text{Slope at } A = \frac{1}{2}\frac{Wl^2}{EI} \text{ radians}$$

(b) *Simply supported beam length l with point load W in middle.*

FOR BENDING MOMENT

at any point:

$$M = \frac{1}{2} Wx$$

$$\text{Max } M = \frac{Wl}{4}$$

FOR DEFLECTION

at any point:

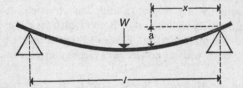

Figure 4.

$$a = \frac{1}{48} \frac{W}{EI} (3l^2 x - 4x^3)$$

$$\text{Max } a = \frac{1}{48} \frac{Wl^3}{EI} \text{ at centre}$$

$$\text{Slope at ends} = \frac{1}{16} \frac{Wl^2}{EI} \text{ radians}$$

(So, *inter alia*, a beam twice as long deflects eight times as much.)

Suggestions for further study

It is not possible to acquire a catholic knowledge of materials entirely by reading books; one must also use direct observation and experience. Just as the real knub of learning photography or drawing or painting lies in learning to see, so one has to cultivate the habit of noticing and observing the structures and materials around one – even the commonest of them – with a seeing eye.

Equally, it is not possible to interpret many of one's observations without the aid of books and, unfortunately, it is not a subject particularly well served by books which are not excessively specialized and mathematical. However, the following books are open to anybody with a knowledge of elementary algebra.

Introduction to Materials Science, by B. R. Schlenker. John Wiley, 1974.

Materials Science (2nd edition), by J. C. Anderson, K. D. Leaver, J. M. Alexander and R. D. Rawlings. Nelson, 1974.

Materials in Industry, by W. J. Patton. Prentice-Hall, 1968.

Structure and Metals, by Michael Hudson. Hutchinson Educational, 1973.

Fibre-Reinforced Materials Technology, by N. J. Parratt. Van Nostrand, 1972.

The Structure and Properties of Materials, Vol. 3, 'Mechanical Behavior', by H. W. Hayden, W. G. Moffatt and J. Wulff. John Wiley, 1965.

Strength of Materials, by Peter Black. Pergamon Press. 1966.

History of the Strength of Materials, by S. P. Timoshenko. McGraw-Hill, 1953.

Strong Solids, by A. Kelly. Oxford University Press, 1966.

The Mechanical Properties of Matter, by A. H. Cottrell. John Wiley, current edition.

Formulas for Stress and Strain, by R. J. Roark. McGraw-Hill, current edition. (This is the Bible of do-it-yourself stressing.)

LESS TECHNICAL BOOKS

Philosophy of Structures, by E. Torroja (translated from the Spanish). University of California Press, Berkeley, 1962.

Metals in the Service of Man, by W. Alexander and A. Street. Penguin Books – current edition.

On Growth and Form, by D'Arcy Thompson (abridged edition). Cambridge University Press, 1961. (This is the classic account of the shape and structure of animals, originally published in 1917 but still readable and important.)

Biomechanics, by R. McNeil Alexander. Chapman and Hall, 1975.

Mechanical Design of Organisms, by S. A. Wainwright, W. D. Biggs, J. D. Currey and J. M. Gosline. Edward Arnold, 1976.

The Southseaman, by J. Weston Martyr. Blackwood, 1928. (An excellent account of classical wooden shipbuilding in the 1920s.)

Engineering Metals and their Alloys, by C. H. Samans. Macmillan, New York, 1953. (This is a good general account of practical metallurgy, rather more advanced than *Metals in the Service of Man*, but very comprehensible and readable.)

Ceramic and Graphite Fibers and Whiskers, by McCreight, Rauch and Sutton. Academic Press, New York, 1965.

Structures or Why Things Don't Fall Down, by J. E. Gordon. Penguin Books, 1978.

Much information about timber can be obtained from the various publications of the Forest Products Research Laboratory which are published by H.M.S.O.

Finally, the *History of Technology* (Oxford University Press, 1954) is a mine of information on all matters connected with the history of materials and structures. This is a forty-guinea five-volume book which is available in many public libraries.

Index

Aircraft:
  weight of, 162, 192
  wooden, 112, 156, 162–72
Akragas (Sicily), 203, 246
Alloy steels (definition), 234
Andrade and Tsien, 81
Ångström unit (definition), 23
Arches, 47–9, 161, 203
Architecture:
  American colonial, 202
  Byzantine, 48, 203
  Gothic, 48–9, 202
  Modern, 264, 267
Asbestos, 120, 270
  in phosphate cement, 175
Atoms, 22, 24, 31, 70, 78, 93, 95, 269
Austenite, 238
Auxins, 132
Avro 504 aircraft, 164

Baekeland, Dr, 179
Banshee, 247
Beam theory, 52–62, 276–9
Beams, 33, 68, 148, 202–4
Beauvais, collapse of, 48
Bernal, J. D., 254
Beryllium, properties of, 195
Bessemer, 251
Bessemer, Sir Henry, 248–52
Bessemer converter, 249–50
Bessemer steel, 247–52
Biggs, Professor W. D., 264
Birkinshaw, John, 246
Blast furnaces, 236, 241, 243, 249

Bligh, Captain, 151
Bonds, chemical (description), 270–72
Bone, 15, 17, 29, 44, 101, 112
Boron fibres, 198–200
Bounty, mutiny in, 151
Brest, blockade of, 150
British Ceramic Research Association, 104
Brittleness, nature of, 101–3
Bronze, 235
Brownian movement, 223, 227
Brunel, I. K., 16, 60, 204

Californian clippers, 151
Cannon, 240–41, 249
Cape Horn, beating round, 151
Carbon fibres, 198–201
Carbon steel, 240, 247
Cars, 191, 263
Cast iron, 42, 44, 45, 139, 240–42, 245, 250
Cellophane, 136
Cellulose, 112, 129–53, 180–83
Cementite, 235–6, 238
Ceramic whiskers, 17, 87–9, 197
Chain anchor cables, 151
Chaplin, Dr C. R., 138
Chitin, 129
Cleopatra's needle, 142
Clocks, grandfather, hardening of gear teeth, 94
Cody, S. F., 163
Cold work, 92, 94, 98
Comet aircraft disasters, 78

Composite materials:
  Cellulose reinforced, 180–83
  Glass reinforced, 183–91, 264
  Moulding powders, 179–80
  Papier-mâché, 177–8
  Primitive, 173
  Reinforced concrete, 202
  Work of fracture of composites, 189
Computers in stress analysis, 113
Concrete, pre-stressed, 111, 204
Cook, John, 113, 114, 118
Cook-Gordon mechanism, 119, 135, 138
Copper sheathing, 152
Cort, Henry, 243
Cost of materials, 260
Cox, H. L., 89
Cracks:
  Griffith, 61, 81, 105, 109, 195
  Stopping by dislocations, 212
  Velocity of propagation, 110, 211
Crash helmets, design of, 104
Creep:
  in metals, 228
  in wood and plastics, 141, 211
Crucible steel, 248
Crystallization, 81, 83, 84, 90, 95, 122, 123, 127, 273
Crystals, regularity of, 95, 225
Cutty Sark, 152

Dash's strong silicon crystal, 90
Democritus, atomic theory of, 66
Dental cements, 112, 175
Devitrification in glass, 83
Diamond, 42, 70, 101, 126
Dislocations:
  diagram of edge, 96
  diagram of screw, 220
  effect of ductility, 95–8, 212–30

  energy of, 222
  mobility of, 95–8, 196, 212–30, 232, 236, 238, 239
  observation of, 225
Disposable technologies, 266
Ductility (definition), 91, 214

E:
  definition, 39–42
  tables of, 42, 193, 194
Elasticity, 18, 19, 30, 33, 35, 42
Elastomers, 42, 73, 275
Elongation, 216
Energy:
  conservation, 262
  definition, 68
  of dislocations, 222
  kinetic, 69
  potential, 68
  strain, 69
  surface, 69–71
Etching of dislocations, 224
Euler collapse, 45, 193
Extraction metallurgy, 231

Faraday, Michael, 15, 66
Fibre saturation point, 145
Fibre-glass, 52, 74, 112, 120, 183–91
Flint, 101, 110, 122
Flutter in aircraft, 193
Folded chain crystals, 127
Forth railway bridge, 247
Forty, Professor J., 220
Frank, Professor F. C., 126, 219
Frank-Read source, 97
Fudge, 247
Fulton's steamship, 231

Galileo, 27, 28, 54
Glass:
  devitrification of, 83

Griffith experiments with, 73–6
theoretical strength of, 74
toughened, 111
Gliders, 129, 140, 166–72
Glucose, 131
Glue:
casein, 156–9
epoxy, 160
hide, 155–6
urea, 159
Glued joints, stress-concentrations in, 157
Grain boundaries, strength of, 90, 250
Griffith, A.A., 67–76, 77
Griffith criterion, 105
Griffith critical crack length, 105
Gurney, Professor Charles, 129

Haematite, 235, 253
Harrer, Heinrich, 122
Hephaistos, 233, 255
Herring and Galt, 85
Hirsch, Professor Sir Peter, 224
Hooke, Robert, 27, 29, 36–8
Hooke's law, 36–8, 71
Hunt, Benjamin, 248
Hydroquinone whiskers, 86, 133
Hydroxyapatite in teeth, 123
Hydroxyl (hydrogen) bonds, 124, 131, 141

Jade, 101, 110, 122
Jeronimidis, Dr Giorgio, 134–5, 138, 190, 201
Joffé, 86

Kaldo process, 252–3
Keller, Professor A., 126–8
Kelly, Dr A., 92, 94
Kelvin, Lord, 22, 229
Kevlar fibre, 199, 201

Lambot, J. L., 204
Laminated wood, 160
Lignin, 137
Lockspeiser, Sir Ben, 73
Locomotives:
cost of, 245
adhesion problems of, 245–6
Lucretius, 66

Magnesia, 67, 196
Mahan, Admiral A. T., 150
Malleability (definition), 215
Manganese in steel, 249–50, 253
Marsh, David, 83–4, 87, 89, 227
Martin, Pierre, 253
Masonry, 46–50, 59, 111, 202–5
Menai railway bridge, 59–60
Menai road bridge, 59–60
Menter, Sir James, 225
Mersenne, Marin, 64
Mica:
Margarite, 122
Muscovite, 120–22
Mice in gliders, 168
Micron (definition), 23
Microscope:
electron, 23, 81, 83, 87, 89, 224, 225
field emission, 24
optical, 23, 81, 89, 223
Mild steel, 247
Mnesicles, 203
Monier, 204
Mosquito aircraft, 165
Moulding powders, 179
Mud huts, Essex, 202
Mushet, Robert, 248–51
Mushet steel, 248

Nelson, Admiral Lord, 150
Newton, Sir Isaac, 28

Oakum, 148
Obsidian, 84, 110, 122, 274
Open-hearth steel, 252
Orowan, Professor E., experiment
　　with mica, 120

Paint, useless to prevent swelling,
　　142
Papier-mâché, 177
Parratt, Margaret, 82
Parthenon:
　chryselephantine statue in, 124
　span of beams in, 203
Pearlite, 238
Pepys, Samuel, 64
Photosynthesis, 130–32, 146
Physical metallurgy, 231
Pig-iron, 239, 240, 243, 249, 252,
　　253
Plant growth, 130
Plasticity, 37, 212, 217
Pliny, 101
Plywood:
　manufacture of, 161
　case-hardening in, 170
Polyethylene, 126–8
Polymers, 124–8, 130–32
Prestressed concrete, 111, 204
Propylea, span of beams in, 203
Pryor, Dr M. G. M., 166–72
Puddled wrought iron, 243
Pyke, Geoffrey, 175

Rabotnov, Professor, 259
Rails:
　cast iron, 244
　wood, 244, 246
　wrought iron, 246
Railways, 27, 40, 60, 141, 244–7,
　　263
Reinforced concrete, 202
Reinforced plastics, 179–201

Roof, air-supported, 52
Rot:
　in gliders, 168
　in timber, 146
Royal Aircraft Establishment,
　　Farnborough, 67, 73, 113,
　　163, 166, 200
Rubber, elasticity of, 42

St Paul, 149
St Paul's Cathedral, 203
St Peter's Cathedral, 203
St Sophia, Constantinople, 48,
　　203
Seasoning of timber, 143
Self-designing structures, 218
Seppings, Sir Robert, 149
Shear strength, estimation of
　　theoretical value, 92
Shot bags in testing, 164
Siemens-Martin steel, 252
Silicon carbide whiskers, 197
Slag, 236, 237, 247, 248, 249,
　　250, 253
Small angle boundaries, 218
Soufflot, J. G., 204
Sound, speed of in solids, 103
Spiegeleisen, 249, 251
Specific Young's modulus, 193
Springs, steam, 246
Stalin, 256
Steam-bending of timber, 143
Steel (definition), 243
Steelmaking, 247
Stephenson, George, 245, 246
Stephenson, Robert, 59, 245
Strength:
　of brittle crystals, 84
　compressive, 45, 192
　tensile, table of, 44
　theoretical, 67–76
　of whiskers, 84–91

Steps, stress concentrations at, 89
Stress concentrations, 19, 77–91, 109, 113–24, 138, 158, 213
Stress corrosion, 214
Stress and strain (definitions), 33–5
Structures, tension and compression, 46
Submarines, strains in hull, 51
Sulphur in steel, 249, 250, 253
Superstitions about materials, 20–22
Surface energy, 68–76, 90, 154
Swords, Japanese, quenched in prisoners, 21

Taylor, Professor Sir Geoffrey, 95–7, 218, 225
Teeth, structure of, 112, 123
Telford, Thomas, 59
Temperature resistance:
  metals, 228, 240
  thermoplastics, 125
  wood, 143
Tempering, 240
Testing machines, 65, 87, 226, 229
Toffee, 105, 210, 274
Tompion, Thomas, 36
Toughness, 15, 101–28, 135, 173–97, 209–30
Toulon, blockade of, 150
Treacle, 131, 274
Trevithick, Robert, 245
Tunicates, 129

Urine, quenching in, 238

Venetian red, 235
Victory, 152, 240–41
Viscosity in toffee and pitch, 105, 210

Wakefield, Edward Gibbon, 261
Wayland Smith, 255
Weston Martyr, J., 64
Whiskers:
  ceramic, 17, 87–9, 197
  crystal, 17, 84–91, 197
  hydroquinone, 86, 133
  metal, 85
Wilkinson, W. B., 204
Wolf, 61
Wood:
  mechanical properties of, 42, 44, 137
  noises in, 140, 142
  optical birefringence, 138
  seasoning of, 143
  swelling of, 141, 148, 162
Wooden ships, 21, 146–53
  leakage in, 21, 148–53
Work of fracture, 108–11
  of composites, 189, 191, 198, 200
  of glass, 108
  of metals, 108, 222
  of thermoplastics, 128
  of wood, 135, 137
Wrought iron, 44, 59, 237, 243, 246

Young's modulus:
  definition of, 39
  tables of, 42, 187, 193, 194, 199
Young, Thomas, 39